Graeme
Feb '7

# LOGIC OF
STATISTICAL INFERENCE

# LOGIC OF STATISTICAL INFERENCE

BY

IAN HACKING

*University Lecturer in Philosophy and
Fellow of Peterhouse, Cambridge*

CAMBRIDGE UNIVERSITY PRESS

Published by the Syndics of the Cambridge University Press
Bentley House, 200 Euston Road, London NW1 2DB
American Branch: 32 East 57th Street, New York, N.Y.10022

© Cambridge University Press 1965

Library of Congress Catalogue Card Number: 66-10043

ISBN: 0 521 05165 7

First published 1965
Reprinted 1974

First printed in Great Britain at the University Printing House, Cambridge
Reprinted in Great Britain by
REDWOOD BURN LIMITED
Trowbridge & Esher

# PREFACE

This book analyses, from the point of view of a philosophical logician, the patterns of statistical inference which have become possible in this century. Logic has traditionally been the science of inference, but although a number of distinguished logicians have contributed under the head of probability, few have studied the actual inferences made by statisticians, or considered the problems specific to statistics. Much recent work has seemed unrelated to practical issues, and is sometimes veiled in a symbolism inscrutable to anyone not educated in the art of reading it. The present study is, in contrast, very much tied to current problems in statistics; it has avoided abstract symbolic systems because the subject seems too young and unstable to make them profitable. I have tried to discover the simple principles which underlie modern work in statistics, and to test them both at a philosophical level and in terms of their practical consequences. Technicalities are kept to a minimum.

It will be evident how many of my ideas come from Sir Ronald Fisher. Since much discussion of statistics has been coloured by purely personal loyalties, it may be worth recording that in my ignorance I knew nothing of Fisher before his death and have been persuaded to the truth of some of his more controversial doctrines only by piecing together the thought in his elliptic publications. My next debt is to Sir Harold Jeffreys, whose *Theory of Probability* remains the finest application of a philosophical understanding to the inferences made in statistics. At a more personal level, it is pleasant to thank the Master and Fellows of Peterhouse, Cambridge, who have provided and guarded the leisure in which to write. I have also been glad of a seminar consisting of Peter Bell, Jonathan Bennett, James Cargile and Timothy Smiley, who, jointly and individually, have helped to correct a great many errors. Finally I am grateful to R. B. Braithwaite for his careful study of the penultimate manuscript, and to David Miller for proof-reading.

Much of chapter IV has appeared in the *Proceedings of the Aristotelian Society* for 1963–4, and is reprinted by kind permission of the Committee. The editor of the *British Journal for the Philosophy of Science* has authorized republication of some parts of my paper 'On the Foundations of Statistics', from volume XV.

I.M.H.

*Vancouver*
*September 1964*

# CONTENTS

*Preface* *page* v

I  LONG RUN FREQUENCIES  1

Statistical inference is chiefly concerned with a physical property, which may be indicated by the name *long run frequency*. The property has never been well defined. Because there are some reasons for denying that it is a physical property at all, its definition is one of the hardest of conceptual problems about statistical inference—and it is taken as the central problem of this book.

II  THE CHANCE SET-UP  13

The long run frequency of an outcome on trials of some kind is a property of the *chance set-up* on which the trials are conducted, and which may be an experimental arrangement or some other part of the world. What the long run frequency is, or was, or would have been is to be called the *chance* of the outcome on trials of a given kind. Chance is a dispositional property: it relates to long run frequency much as the frangibility of a wine glass relates to whether the glass does, or did, or will break when dropped. The important notion of a distribution of chances is defined, and chances are shown to satisfy the well-known Kolmogoroff axioms for probability. These are the beginnings of a postulational definition of chance.

III  SUPPORT  27

Although the Kolmogoroff axioms help to define chance they are not enough. They do not determine when, for instance, an hypothesis about chances—a statistical hypothesis—is well supported by statistical data. Hence we shall need further postulates to define chances and to provide foundations for inferences about them. As a preliminary, the idea of support by data is discussed, and axioms, originally due to Koopman, are stated. Much weaker than the axioms commonly used in statistics, they will serve as part of the underlying logic for further investigation of chance.

IV  THE LONG RUN  39

Some connexions must be established between chance and support. The simplest is as follows: given that on some kind of trial an event of kind $A$ happens more often than one of kind $B$, then the proposition that $A$

occurs on some individual trial is, lacking other data, better supported than the proposition that $B$ occurs at that trial. Something like this feeble connexion is widely held to follow from its long run success when used as the basis for a guessing policy. But no such long run defence is valid, and the connexion, if it exists at all, must have an entirely different foundation.

## V  THE LAW OF LIKELIHOOD 54

A principle named the *law of likelihood* is proposed as an explication of the banal connexion stated in the preceding chapter, and then is used as a postulate to be added to Koopman's and Kolmogoroff's axioms as part of the postulational definition of chance. It relies on mere comparisons of support, justifying conclusions of the form, 'the data support this hypothesis better than that'. But it is relevant not only to guessing what will happen on the basis of known frequencies, but also to guessing frequencies on the basis of observed experimental results. Any such suggested law must be regarded as a conjecture to be tested in terms of its consequences, and this work is commenced.

## VI  STATISTICAL TESTS 74

The traditional problem of testing statistical hypotheses is examined with a view to discovering general requirements for any theory of testing. Then it is shown how one plausible theory is in exact agreement with the law of likelihood.

## VII  THEORIES OF TESTING 89

A rigorous theory of testing statistical hypotheses is developed from the law of likelihood, and other theories are contrasted with it. The Neyman-Pearson theory is given special attention, and turns out to be valid in a much narrower domain than is commonly supposed. When it is valid, it is actually included in the likelihood theory.

## VIII  RANDOM SAMPLING 118

The theory of the preceding chapters is applied to inferences from sample to population. Some paradoxes about randomness are resolved.

## IX  THE FIDUCIAL ARGUMENT 133

So far in the essay there has been no way to measure the degree to which a body of data supports an hypothesis. So the law of likelihood is strengthened to form a *principle of irrelevance* which, when added to other standard axioms, provides a measure of support by data. This measure may be regarded as the consistent explication of Fisher's hitherto inconsistent theory of fiducial probability; at the same time it extends the postulational definition of chance within the pre-established underlying logic. It is proved that any principle similar to the principle of irrelevance, but stronger than it, must lead to contradiction. So the principle completes a theory of statistical support.

## X  ESTIMATION                                                                          161

A very general account of estimation is provided in preparation for a special study of estimation in statistics.

## XI  POINT ESTIMATION                                                        174

The traditional theory of point estimation is developed, so far as is possible, along the lines of the preceding chapter, and relying on the theory of statistical support. Unfortunately the very concept of estimation seems ill adapted to statistics, and unless other notions are imported, it is impossible to define a 'best estimate' for typical problems. At most *admissible estimates* can be defined, which are not demonstrably worse than other possible estimates. Usually the currently popular theories of estimation provide estimates which are admissible in the sense of the present chapter, but where popular estimates diverge, there is seldom any way of settling which is best without some element of arbitrary convention or whim.

## XII  BAYES' THEORY                                                              190

Because of the occasional limitations in the theory of statistical support, the bolder theories of Bayes and Jeffreys are examined, but each is rejected for reasons which by now are entirely standard.

## XIII  THE SUBJECTIVE THEORY                                          208

Bayes' ideas have naturally led to a subjective theory of statistical inference, which is here explained sympathetically, and shown to be in principle consistent with our theory of statistical support. Neo-Bayesian work claims to analyse a much wider range of inferences than our theory attempts, but ours gives the more detailed account of the inferences within its domain, and hence it has the virtue of being more readily open to refutation and subsequent improvement.

*Index*                                                                                              229

CHAPTER I

# LONG RUN FREQUENCIES

The problem of the foundation of statistics is to state a set of principles which entail the validity of all correct statistical inference, and which do not imply that any fallacious inference is valid. Much statistical inference is concerned with a special kind of property, and a good deal of the foundations depends upon its definition. Since no current definition is adequate, the next several chapters will present a better one.

Among familiar examples of the crucial property, a coin and tossing device can be so made that, in the long run, the frequency with which the coin falls heads when tossed is about 3/4. Overall, in the long run, the frequency of traffic accidents on foggy nights in a great city is pretty constant. More than 95% of a marksman's shots hit the bull's eye. No one can doubt that these frequencies, fractions, ratios, and proportions indicate physical characteristics of some parts of the world. Road safety programmes and target practice alike assume the frequencies are open to controlled experiment. If there are sceptics who insist that the frequency in the long run with which the coin falls heads is no property of anything, they have this much right on their side: the property has never been clearly defined. It is a serious conceptual problem, to define it.

The property need not be static. It is the key to many dynamic studies. In an epidemic the frequency with which citizens become infected may be a function of the number ill at the time, so that knowledge of this function would help to chart future ravages of the disease. Since the frequency is changing, we must consider frequencies over a fairly short period of time; perhaps it may even be correct to consider instantaneous frequencies but such a paradoxical conception must await further analysis.

First the property needs a name. We might speak of the ratio, proportion, fraction or percentage of heads obtained in coin tossing, but each of these words suggests a ratio within a closed class. It is important to convey the fact that whenever the coin is tossed sufficiently often, the frequency of heads is about 3/4. So we shall say, for the present, that the *long run frequency* is about 3/4. This is only a label, but perhaps a natural one. It is hardly fitting for epidemics, where there is no long run at constant frequency of infection because the frequency is always changing. Better terms will be devised presently. It is easiest to begin with cases in which it does make sense to speak of a long run. In the end, the whole of this vague notion, the long run, must be analysed away, but in the beginning a constant reminder of its obscurity will do no harm.

The long run frequency of something is a quantity. I have called long run frequency a property, as one might call density and length properties. That convenience relies on an understood ellipsis. It is not length which is a property of a bar of iron; rather a particular length, the length of the bar, is a property of the bar. Likewise long run frequency is not a property of any part of the world, but a particular long run frequency of something may be a property of some part of the world. In what follows I shall use the word 'frequency' in connexion with this property only, and not, for example, to denote a proportion in a closed class of events.

Long run frequency is at the core of statistical reasoning. Hence the forthcoming analysis will continually employ the discoveries of statisticians, who are the people who have thought most about it. Some statisticians use the word 'probability' as the name of the physical property I have in mind, and never use that word in any other way. Others sometimes so use it and sometimes not; a few never use it to name a property. But however you think the word ought to be used, there is no denying that some statisticians do use it as the name of the property I label long run frequency. In what follows I shall not so use the word; in fact I shall scarcely use the word 'probability' at all. There is nothing wrong with using it to name a physical property if one makes plain what one is doing, but I avoid this practice to circumvent mere verbal controversy.

Most statisticians and probability theorists have no qualms over calling long run frequency a property. One of the most eminent says,

Let the *frequency* of an outcome $A$ in $n$ repeated trials be the ratio $n_A/n$ of the number $n_A$ of occurrences of $A$ to the total number $n$ of trials. If, in repeating a trial a large number of times, the observed frequencies of any one of its outcomes $A$ cluster about some number, the trial is then said to be *random*. For example, in a game of dice (two homogeneous ones) 'double-six' occurs about once in 36 times, that is, its observed frequencies cluster about 1/36. The number 1/36 is a permanent numerical property of 'double-six' under the conditions of the game, and the observed frequencies are to be thought of as measurements of the property. This is analogous to stating that, say, a bar at a fixed temperature has a permanent numerical property called its 'length' about which the measurements cluster.†

The author states that only recently have men studied this property. If someone notices a new property like this, a definition might not be needed. Perhaps even in rigorous work it might suffice to point out the property by examples and vague description, to name it, and proceed at once to investigate empirical laws in which it occurs. This will do only if the property is readily recognized, is formally much like other properties, and seems to have no peculiarities of its own. But frequency in the long run is very peculiar.

The analogy between length and frequency in the long run seems strained. Suppose, for example, that a bar is measured by a meter stick, and that the measurements cluster about 25 cm. Any measurement over 50 cm. is either wildly wrong, or was made under extraordinary conditions. Within current working theories there is no alternative. But if one measurement of double-six is 2/36, while the average is 1/36, there seems no reason to suspect error or changing conditions. There might be good reason if 2/36 were based on a very long sequence of trials. There's the well-known rub. How long? 'Many successive trials', and a 'large number of times' are all too vague. So we demand a fuller definition of this alleged physical property, long run frequency.

It might be said here that the analogy between length and frequency in the long run does not break down; there are measurements on dice tossing just as indicative of error as measurements of length. As is well known, the theory of measurements is nowadays a part of the study of frequencies. But a definition or at least deeper analysis is required before it is evident that frequency in the long run is a well-behaved property at all. Only after its analysis can it be glibly compared to more familiar things.

† Michel Loève, *Probability Theory* (New York, 1955), p. 5.

*Definition*

A definition of frequency is needed. The definition of a word of English is primarily a matter of linguistics, while to define an arbitrarily introduced new term may be pure stipulation. My task is neither of these, but to define a property; to draw, as it were, a line around the property to which the examples point. When, on the basis of experiment, can one conclude that something has this property? When are such hypotheses refuted by evidence, and when well supported by it?

A definition might be a form of words equivalent in meaning to the name of the property one has in mind. But that almost certainly can't be given here. There is another excellent way of defining; to state several facts about this property by way of postulates. Such a definition is for practical purposes complete if every correct inference essentially involving this property is validated by the postulates, without recourse to any unstated facts about the property—while at the same time no incorrect inference is authorized.

Such a postulational definition of frequency in the long run must not only cover what have been called direct inferences: inferences from the fact that something has a frequency of $x$ to something else having a frequency of $y$. It must also cover what have been given the rather odd name of inverse inferences: from experimental data not mentioning frequency in the long run, to frequency in the long run. It has sometimes been suggested that a complete definition of the property need not cover the second kind of inference. This is absurd; a definition which does not authorize any inferring from experimental data to good support for the proposition that something has, or has not, the property in question, cannot be called a definition of an empirically significant property at all.

At one time many practical consequences might have followed a complete postulational definition. There need not be many today. The property is pretty well understood by those who use it. Their descriptions of it seldom seem completely apt, but their work with it includes some of the most remarkable discoveries made in any science. Hence our task resides squarely in the philosophy of science: to understand this great work rather than to improve it.

## DEFINITION

Statistical results will be used throughout the rest of this essay, which is little more than a philosophical codification of things already known.

*Frequency in an infinite sequence*

One definition of frequency in the long run is especially celebrated. 'The fundamental conception which the reader has to fix in his mind as clearly as possible is', according to Venn, 'that of a series'; he continues by describing a special kind of series which 'combines individual irregularity with aggregate regularity'.† At least part of his book may be construed as analysing a property of this kind of series. It gave expression to an idea which was much in the air at the time. Cournot, Ellis and others had written in the same vein a good deal earlier.

Von Mises refined the idea; his work justly remains the classic exposition of the theory. He studied a kind of series satisfying certain conditions, and which he called a collective. Von Mises' conditions have been modified; the modified conditions have been proved consistent. For any kind of event $E$, and collective $K$, consider the proportion of times in which events of this kind occur in the first $n$ members of $K$. Call this $P_n(E)$. If $P_n(E)$ approaches a limit as $n$ becomes large without bound, represent the limit by $P(E)$. This is a property of the collective $K$. Presumably it is a property of some part of the world that it would generate a collective with certain $P(E)$ for some $E$. $P(E)$ is von Mises' explication of frequency in the long run.

Is this property $P(E)$ empirically significant? Could the proposition that something has the property $P(E)$ in collective $K$ ever be supported by experimental data? The questions have been asked repeatedly since Mises first published. Suppose it were granted or assumed that an experimental set-up would generate a collective. Could any data indicate the $P(E)$ in the collective? Presumably the outcome of a long sequence of actual experiments should be considered an initial or early segment of the collective. But any finite segment is compatible with, and does not give any shred of indication of, any limiting value whatsoever. On the basis of data about a finite segment, the hypothesis that a limiting value is 9/10 is, as far as experimental support is concerned, on a par with the

† John Venn, *The Logic of Chance* (London and Cambridge, 1866), p. 4.

claim that the limit is 1/10, or the suggestion that there is no limit at all. Other principles are needed than mere analyses of limits.

This standard objection is not just that no propositions about the $P(E)$ in $K$ are conclusively verifiable or refutable in the light of experience. The trouble is that no amount of experience can, in the literal terms of the theory, give any indication whatever of the limiting value. Nor, if 'limiting value' be taken literally, is there any reason for saying observed proportions even approach a limiting value.

Von Mises' theory could be supplemented by extra principles showing how data might support, vindicate, or refute an hypothesis about the limit. But these would not be optional extras: without them there is no experimentally significant property at all. Of course von Mises' theory can be supplemented in this way. Reichenbach is one of several who attempts it. But I shall argue that the simplest principles which are adequate are also adequate for defining frequency even if frequency is not conceived as the limit of an infinite series. Infinity is redundant.

Probably von Mises did not intend his theory to be taken quite as I imply. He wished to present a mathematical conception which was in some way an idealization of the property which I label frequency in the long run. But then there remains the problem of defining that which he has idealized. Nor is it obvious that von Mises' idealization is very useful. It is sometimes said that the Euclidean plane or spherical geometry used in surveying involves an idealization. Perhaps this means that surveyors take a measurement, make simplifying assumptions, use Euclid for computing their consequences, and finally assume that these consequences are also, approximately, properties of the world from which the original measurements were taken.

It is true that some of the laws of von Mises' collective apply to frequency in the long run, and that these laws are used in computing new frequencies from old. But it is the laws, and not the infinite collective, which are of use here. Never, in the journals, will one find a statistician using the peculiar characteristics of a collective in making a statistical inference, whereas surveyors really do use some of the attributes peculiar to Euclidean plane or spherical geometry. So whatever its interest in its own right, the theory of collectives seems redundant as an idealization in the study of frequency in the long run.

Von Mises may feel that what is required from a philosophical point of view is not a mere statement of laws used by statisticians, but a coherent idealization which knits together all their principles. He might admit that you can get along without the idealization, but deny that you can see the relationship between the principles without using it. However sound may be this idea, von Mises' own theory fails to satisfy it. For the crucial principles in statistics concern the measurement of our physical property. There must be some principles for inferring frequencies from experimental data: and they are the ones which make frequency interesting. An idealization which lacks these principles fails where it is most needed.

There is no denying the intuitive appeal of replacing frequency in the long run by frequency in an infinite run. This old idea has stimulated much statistical imagination. For several decades of this century Fisher was the most prolific genius of theoretical statistics. In much of his work he refers to properties of 'hypothetical infinite populations'. These populations are at only one remove from von Mises' collectives; von Mises' work can even be construed as assigning them logical precision. However much they have been a help, I shall argue that hypothetical infinite populations only hinder full understanding of the very property von Mises and Fisher did so much to elucidate.

*Axiomatic models*

Most modern workers in theoretical statistics differ from von Mises. Their theory may be expressed as follows. There are various phenomena, and a property, which can be sketched by examples and vague description. The phenomena are 'random', the property, 'frequency in the long run'. The business of the formal theory of probability is to give a mathematical model of these phenomena. 'Probability' is the name given to anything satisfying the axioms of this model. Aside from some intuitive words of explanation, that is all that needs to be said about these phenomena.

This attitude is perfectly correct if one is concerned with what is now generally called probability theory, which makes deductions from a few axioms, regardless of their interpretation. But it is more dubious in statistics, which uses a unique interpretation of those axioms. Insistence on certain definable mathematical properties is of course salutary. The axioms, in a form due to Kolmogoroff, are indeed an essential tool of statistics. Frequency in the long run does, as we shall show, satisfy them: they are part of the postulational definition of frequency in the long run.

The formal theory based on Kolmogoroff's axioms is the means for proving conditional propositions, 'if the frequency in the long run of something is *this*, then that of something else is *that*'. The theory is also a superbly rich abstract discipline, one of the half dozen most stimulating fields of pure mathematics today. But pointing out examples and presenting the formal theory does not provide a complete definition of frequency in the long run. It only provides results which will assist the definition. From no theorem of the formal theory can one infer that any hypothesis about frequency in the long run is, for instance, well supported by certain experimental data. Other postulates are needed; until they are given it is not evident that any experimental property is well defined. The forthcoming work will add other postulates to those of Kolmogoroff, in the hope of completing a definition of our property.

Text-books persistently repeat the idea that the formal theory conveyed by Kolmogoroff's axioms is a model for frequency in the long run. If this does not conceal an actual confusion, it at least makes it easy to ignore problems about long run frequency. A formal mathematical theory consists essentially of vocabulary, syntax, and a set of axioms and rules of proof, all of which may be embedded in some commonly accepted logic. A theory may be modelled in another, but that does not concern us here. A science, or part of a science, is a model for a formal mathematical theory when there is a mapping between sentences of the theory, and propositions germane to the science, such that those sentences of the theory which can be derived from the axioms map on to true propositions from the science. Conversely, the theory is a model of the science when truths of the science map on to the theorems of the theory. The assertion that one is a model of the other demands, as far as I can see, that the science have clear-cut intelligible propositions. To put the case as boldly as possible, the science of long run frequency has not been proved to have such propositions. More cautiously, no clear-cut account has ever been given of their meaning or logic of verification. To call the formal theory a model of the science, and to say a term in that theory models the property of frequency in the long run, is to beg the questions at issue: are there clear-cut propositions of the science? And, how is the property to be defined?

Von Mises is plain on this very issue. What is called the formal theory of probability is a part of set theory, being the study of a special class of measures on sets of points. Referring to the property which I call frequency in the long run, von Mises insists that the theory of it 'can never become a part of the theory of sets. It remains a natural science, a theory of certain observable phenomena, which we have idealized in the concept of a collective'.

*Another kind of model*

To avoid confusion, it is worth mentioning a quite unexceptionable use of the word 'model' in studying frequency. It is often helpful and sometimes essential to make radical simplifying assumptions about the structure of a process. The whole of economics uses such assumptions all the time, and it is said to make models, in this sense, of coal-consumption in underdeveloped countries or of soap-production in developed ones.

There are many well-known frequency models of, for example, the transmission of a contagious disease. Suppose that every new infection increases the frequency with which further infection occurs. A model of this situation consists of an urn with $b$ black and $r$ red balls. Balls are drawn and replaced; if a red ball is drawn (a new infection) $c$ more red balls are added also (increasing the frequency with which red balls are drawn). The urn is shaken after each replacement.† The urn provides a model of an epidemic, and so do the mathematical laws governing the behaviour of the urn. But no one would suggest that without further postulates one could ever know, for instance, how good a simplification one had made. Such models are a rich source of discoveries. But they do not have much to do with the very definition of the property they employ.

† The idea is due to G. Polya; see William Feller, *An Introduction to Probability Theory and its Applications* (New York, 1950), p. 109.

## Braithwaite's theory

There is one philosophical theory nearer than any other to my ideal of postulational definition of frequency in the long run. Braithwaite proposes to state the meaning of assertions about this property in terms of rules for rejecting them on the basis of experimental data.† If these rules be added to Kolmogoroff's axioms, one has a more stringent set of postulates than the axioms alone, and which do bear on the experimental application of the property. I am certain his rules will not do, and will argue this in due course. For the present a milder remark will suffice: they are certainly not complete. For it will typically happen that on any experimental data whatsoever, a large class of statistical hypotheses will not be rejected by Braithwaite's rules. Now on the same data it would generally be admitted that some hypotheses are better supported than others. That they are better supported presumably follows, in part, from the logic of our property, frequency in the long run. No definition from which this does not follow can be wholly adequate. It does not follow from Braithwaite's rules, taken together with Kolmogoroff's axioms.

## Chance

It has been convenient, in this introductory chapter, to speak of a property and label it with the phrase, 'frequency in the long run'. But already it is apparent that this term will not do. 'Frequency in the long run' is all very well, but it is a property of the coin and tossing device, not only that, in the long run, heads fall more often than tails, but also that this would happen even if in fact the device were dismantled and the coin melted. This is a dispositional property of the coin: what the long run frequency is or would be or would have been. Popper calls it a propensity of the coin, device, and situation. Now if a wine glass would break, or would have broken, or will break, when dropped, we say the glass is fragile. There is a word for the active event, and another for the passive dispositional property. It will be convenient to have a plainly dispositional word for our property—a brief way of saying what the long run frequency is or was or would have been. 'Probability' is often so used, but I eschew it here. So I shall resurrect a good

† R. B. Braithwaite, *Scientific Explanation* (Cambridge, 1953), ch. vi.

seventeenth-century word which seems sometimes to have been used in just this sense—*chance*. Jeffreys uses the word in much the same way, and so does Kneale.

So when I say the chance of getting heads on a toss of this coin from a certain device is such and such, I refer to what the frequency in the long run is or would have been (etc.). No more. Our task is to define this property. So far it has only been indicated.

I shall still sometimes refer to the physical property I have in mind as the frequency in the long run. This may seem a less appropriate way of speaking, but it will, for the next few chapters, be a useful reminder of what we have to define. I have rather arbitrarily chosen 'chance' to denote our property; there is always a danger of ignoring the allocated sense or simply forgetting it. So an occasional reminder in terms of frequency in the long run will do no harm.

'Chance' is a convenient word. Take the urn model of an epidemic. The urn contains $b$ black balls and $r$ red ones; a trial consists of drawing a ball and noting its colour. The ball is replaced; if it is red, $c$ red balls are added too. All are shaken and another trial is made. It often happens with urns that the long run frequency of drawing red is equal to the proportion of balls in the urn which are red. At any stage in the sequence of trials it makes good sense to ask what the frequency in the long run would be, if we began drawing and replacing balls without adding any more red ones. That is the chance of red at a particular point in the sequence of trials—namely what the frequency would have been if.... This 'would have been' is Pickwickian when the true epidemic behaves like the urn. At any point, the chance of a new infection is 'what the frequency in the long run of new infections would have been, if only new infections did not alter the chance of infection'. This verges on nonsense. It at best conveys a picture. What we need are postulates about chance which conform to this picture, but which are absolutely rigorous and not for ever bound up with the long run.

Untidy examples like this are useful now. It is unlikely that the logic of what the frequency in the long run would have been is very different from the logic of static long run frequencies exemplified in coin tossing. This is especially clear if one thinks of the urn model of the epidemic. Yet not only is there no operational infinite long run for the real epidemic, there is no run of any sizeable length at all, since in the beleaguered city the chance of infection varies from day to day. Of course our devising a dispositional sounding word like chance in no way shows that the property can be defined for the dynamic cases, but an adaptable terminology will make the definition a little easier.

CHAPTER II

# THE CHANCE SET-UP

Of what kind of thing is chance, or frequency in the long run, a property? Early writers may have conceived chances as properties of things like dice. Von Mises defines probability as the property of a sequence, while Neyman applies it to sets called fundamental probability sets. Fisher has an hypothetical infinite population in mind. But a more naïve answer stands out. The frequency in the long run of heads from a coin tossing device seems to be a property of the coin and device; the frequency in the long run of accidents on a stretch of highway seems to be a property of, in part, the road and those who drive upon it. We have no general name in English for this sort of thing. I shall use 'chance set-up'. We also need a term corresponding to the toss of the coin and observing the outcome, and equally to the passage of a day on which an accident may occur. For three centuries the word 'trial' has been used in this sense, and I shall adopt it.

A *chance set-up* is a device or part of the world on which might be conducted one or more *trials*, experiments, or observations; each trial must have a unique *result* which is a member of a *class of possible results*.

A piece of radium together with a recording mechanism might constitute a chance set-up. One possible trial consists in observing whether or not the radium emits radiation in a small time interval. Possible results are 'radiation' and 'none'. A pair of mice may provide a chance set-up, the trial being mating and the possible results the possible genetic make-ups of the offspring. The notion of a chance set-up is as old as the study of frequency. For Cournot, frequencies are properties of parts of the world, though he is careful not to say exactly what parts, in general. Venn's descriptions make it plain that he has a chance set-up in mind, and that it is this which leads him to the idea of an unending series of trials. Von Mises' probability is a property of a series, but it is intended as a model of a property of what he calls an experimental set-up— I have copied the very word 'set-up' from his English translators.

More recently, Popper has urged that the dispositional property of frequency in the long run is a property of an experimental 'arrangement'.

The same physical object might be the basis of many different kinds of trial. For the coin tossing device, so made that the coin must fall either heads or tails, one might:

(1) Toss up and note result. There are 2 possible results, heads and tails.

(2) Toss thrice and note number of $H$. 4 possible results.

(3) Toss thrice and note actual arrangement of $H$ and $T$ among the outcomes. 8 possible results.

(4) Toss once and note whether the coin falls on edge or not. 1 possible result.

(5) Toss once and abandon trial if the coin falls $T$, otherwise, toss again and note the result. 2 possible results.

(6) Toss once and abandon trial if coin falls $H$ or $T$; if not, toss again, and note the result. 0 possible results.

Case (4) is degenerate, and (6) is a trial which can never be completed. Cases (5) and (6) are *conditional* trials, whose result is conditional on another trial having a certain result. Notice that essentially the same event may provide many different kinds of trial. One kind of trial when the coin is tossed thrice has 4 possible results, namely the number of heads. Another has 8, namely the exact order of $H$ and $T$.

It will be much easier to discuss sets of results rather than individual ones. Since each trial must have a unique result, it is assumed that the possible results for any trial are mutually exclusive. A set of possible results on any trial will be called an *outcome*. A trial will be said to have outcome $E$ when the result is a member of $E$.

The *union* of two sets $E$ and $F$ is the set which contains just those things which are either in $E$ or in $F$. The *intersection* is the set containing those things which are in both $E$ and $F$. The union is written $E \cup F$, and the intersection, $E \cap F$. Outcomes are sets of results. Hence a trial has outcome $E \cup F$ if the result of the trial is in either $E$ or $F$; it has outcome $E \cap F$ if the result is in both $E$ and $F$.

Corresponding to each result is that outcome whose only member is that result, and which in set theory would be called the *unit set* of that result. In this book, results and their unit sets will be distinguished only when really necessary. Sometimes an outcome with only one member will be referred to as a result. This procedure is quite innocent. In fact one of the most famous of logical systems, one due to Quine, explicitly and deliberately equates things and their unit sets.

Finally, to complete this indispensable battery of terms, the set of all possible results of a trial of given kind will be called the sure outcome, represented by $\Omega$. Every trial must have outcome $\Omega$.

When the occurrence of trials on a chance set-up is itself ordered in a law-like way, the whole will be called a *chance process*. It is of special interest if the chance of an outcome on one trial should depend on the outcome of preceding trials. In its infancy, statistics dealt almost entirely with static chance set-ups. Now it is increasingly involved in chance processes, often called stochastic processes. It is currently supposed that the world is a chance process.

*Random variables*

I am sorry I cannot always use terms which are customary in statistics. A philosophical essay about statistical science ought to use the words already current. But we cannot safely use 'probability' for 'chance' because the word has been employed in too many different ways. In the course of this essay we shall need to refer to practically everything which the word 'probability' has ever been used to refer to; to avoid confusion it is better to abandon the word altogether and use other expressions which have not yet been made so equivocal.

A more troublesome deviation from standard practice concerns my talk of kinds of trial. It is, I think, quite straightforward, but statisticians usually call the outcome of a trial of some kind a *random variable*. A number of writers have already objected that this use of the word 'variable' is at odds with others common in mathematics.† Still more suspicious is the adjective 'random'. We want to define randomness in due course, but not to take it for granted in a primitive term. A random variable, apparently, is

† K. Menger, 'Random variables from the point of view of a general theory of variables', *Third Berkeley Symposium on Probability and Statistics* (Berkeley and Los Angeles, 1956), II, pp. 215-29.

something which varies randomly, but what is that? It is possible entirely to avoid the language of random variables, and although in this way we shall sacrifice some glib brevity, fundamental difficulties should be made more clear.

*Distributions*

Venn called the series 'the fundamental conception which the reader has to fix in his mind'. Today, perhaps, the *distribution of chances* is the fundamental notion. It can be explained regardless of the precise sense which may be given to chance, or frequency in the long run.

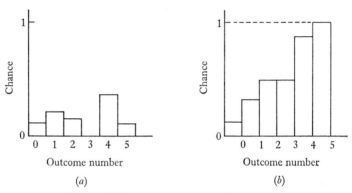

Fig. 1. (*a*) Bar graph. (*b*) Cumulative graph.

A chance set-up is partially described by stating the chance of one outcome on some kind of trial; a more comprehensive description states the way in which the chances are distributed amongst the various outcomes. Instead of stating the chance of 3 heads in three tosses, one may state the chances for each possible result.

There are two distinct ways of presenting a distribution. I begin with cases in which there are finitely many possible results. The chances of each might be presented by a bar graph, or histogram, like that on the left. But let the results be numbered, 0, 1, 2, ..., *n*. Exactly the same information as that conveyed by the bar graph may be conveyed by a step graph, like that on the right. This shows the chance of getting a result numbered not more than 0, not more than 1, not more than 2, and so on. The chance of getting a result not more than 3 is, of course, just the chance of getting a result numbered not more than 2, plus the chance of getting result 3. The graph on the right shows how the chances accumulate and so is called cumulative. The curves of the two graphs may be represented by mathematical functions; the curve of the first graph will simply be called a *bar function*, that of the second, a *cumulative function*. They are merely alternative ways of describing the same thing. If the bar function is $f$, and $f(2) = 0.6$, the chance of result 2 is 0.6. If the cumulative function is $F$, and $F(2) = 0.4$, the chance of getting a result numbered 2 or less is 0.4. Evidently $F(n) = F(n-1) + f(n)$, or again, $F(n) = f(0) + f(1) + \ldots + f(n)$.

This is of little interest except by way of a very elementary, intuitive introduction to continuous distributions, in which there are more than finitely many possible outcomes of a trial on a set-up. Consider the phone calls arriving at a central exchange in a peak period. The *waiting time* between calls is the time interval between the arrival of one incoming call and the next. The trials might consist of noting of each successive waiting time whether it is less than a second long, between 1 and 2 sec., and so on. Evidently the distribution of chances for these results can be represented by a bar graph like that above, and also by a step or cumulative graph. But we may consider trials whose results involve smaller intervals, say measured to the nearest tenth or ten-thousandth of a second. Then the chance of the waiting time falling in any particular interval may be tiny. But the corresponding cumulative graph will, as one decreases the size of interval, progress from a bumpy lot of steps to something like a smooth curve. Hence the cumulative function approximates to a curve, and is most naturally specified by a smooth curve, or continuous function. In this way one is led to the idea of a set-up with continuously many possible results; a waiting time may be any real number of seconds from 0 on. Even if there is zero chance that the waiting time is exactly 1 sec. long, the chance that it is less than 1 sec., or less than 10 sec. will still be expressible in terms of a cumulative function.

When the cumulative function is continuous and continuously differentiable, there is a useful analogy to the bar function for discrete distributions. In the discrete case the value of the cumulative $F(n)$ is the sum of the values of $f$ for arguments up to $n$. In the continuous case, for a cumulative function $G$, we consider the function $g$ such that the chance of getting an outcome smaller than

$t_0$—the length $G(t_0)$ on the right-hand diagram—is equal to the area under $g$ up to $t_0$. This area has been shaded on the left-hand sketch; its analytic expression is of course

$$G(t_0) = \int_0^{t_0} g(t)\,dt.$$

$g$ is simply the derivative of $G$ with respect to time. It is called the *density* function for the cumulative $G$. The idea is that $g$ indicates the relative densities of the chances at different possible results.

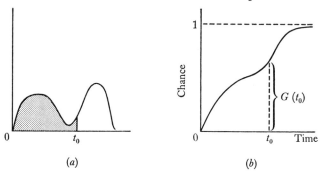

Fig. 2. (*a*) Density function $g$. (*b*) Cumulative function $G$.

For all the abstraction involved in these ideas, the notion of a distribution is the key to the study of chance. It will be shown in the sequel that there is no general method for testing hypotheses about frequency in the long run, which is not specifically a method for testing hypotheses about distributions.

*Kolmogoroff's axioms*

It is easy to agree on some of the laws of chance. They might be introduced in various ways. One brief heuristic argument will be sketched here. But assent to the laws is so universal that the argument will be neither complete nor perfect. It is noteworthy that the laws hold of frequency in the long run under any meaning of the 'long run' ever propounded.

The chance of a certain outcome $E$ on a trial of kind $K$ is a property of a chance set-up. It could be abbreviated in various ways but when the context makes evident which kind of trial is in question, it suffices to write $P(E)$. Strictly speaking we are concerned with an ordered triad: the set-up, the kind of trial, the outcome.

## KOLMOGOROFF'S AXIOMS

Among the possible answers to the question, how frequently? is a range, perhaps expounded in percentages, from 'never' through '1 % of the time' through 'half the time' to '99 %' and 'always'. If we characterize the frequency of that which happens 100 % of the time by 1, and that which never happens by 0, then, for a given kind of trial,

*Axiom 1:* $\qquad 0 \leqslant P(E) \leqslant 1.$

Where $\Omega$ stands for the sure outcome,

*Axiom 2:* $\qquad P(\Omega) = 1.$

Finally, if two outcomes are mutally exclusive, the frequency with which one or the other occurs is just the sum of the frequencies with which each occurs. This is true for any length of run: long, short, or infinite.

*Axiom 3.* If $E$ and $F$ are mutually exclusive,
$$P(E \cup F) = P(E) + P(F).$$

If a trial has infinitely many possible results, let $E_1, E_2, \ldots$ be any countably infinite list of some of its possible outcomes. Then $\cup E_j$ shall stand for the countable union of these outcomes: the set including all results which are members of outcomes on the list. $\Sigma P(E_j)$ is the sum of the chances of the outcomes on the list. We extend axiom 3, insisting that for any countably infinite set of outcomes, $\cup E_j$ is a possible outcome, $\Sigma P(E_j)$ exists, and that if for all $i$ and $j$, $i \neq j$, $E_i$ and $E_j$ are disjoint, then $P(\cup E_j) = \Sigma P(E_j)$.

## Conditional chance

It should be sensible to speak of the chance that a trial has outcome $E$, on condition that it has outcome $F$. This is in fact just the chance of an outcome on a conditional trial. It will be represented by $P(E|F)$. For example, consider trials consisting of two consecutive tosses of a coin. The chance of heads on the second toss, on condition of tails on the first, is just the frequency in the long run with which pairs of tosses whose first member is $T$, have $H$ as second member.

Suppose that in any sequence of $m$ trials, outcome $F$ occurs $n$ times, and that $E \cap F$ occurs $k$ times. A subset of this sequence of $m$ trials is a sequence of conditional trials: trials on condition that the outcome is $F$, and with possible results $E$ and not-$E$. We may compute the frequency of $E$ on condition that $F$. It is just the frequency of $E \cap F$ amongst trials whose outcome is $F$. Hence $k/n$. But if $m \neq 0$, $k/n = (k/m)/(n/m)$. So the frequency of $E$ on condition $F$ equals the frequency of $E \cap F$ divided by that of $F$. If $P(F) > 0$, we define

$$P(E|F) = \frac{P(E \cap F)}{P(F)}.$$

This argument is good for frequencies in any length of run except an infinite one, but the definition is easily shown appropriate there too. So that definition is fitting to conditional chance, whatever is meant by the long run.

*Independence*

Two events are commonly said to be independent of each other if neither causes the other, and if knowledge that one occurred is of no aid in discovering if the other occurred. It might seem as if our work so far assumed that outcomes on successive trials of some kind should be independent. So much is often built into definitions of chance, but our exposition assumes nothing of the sort. Hence it accords with common-sense, for if heads and tails alternate on trials of some kind, then the long run frequency of heads will, I suppose, be 0·5, but outcomes certainly are not independent of each other. Thus chance or long run frequency does not presuppose the notion of independence, and we may guess that independence is to be defined via our primitive terms, *chance* and *kind of trial*.

There seems no way to define the independence of particular events within the framework of long run frequency, but we shall need only the independence of outcomes on trials of some kind. To begin with, take two possible outcomes of trials of kind $K$, say $E_1$ and $E_2$. If these outcomes are independent on trials of this kind, the long run frequency with which $E_1$ occurs among trials yielding $E_2$ should be the same as in trials in general. Otherwise knowledge that $E_2$ occurred at some designated trial would be relevant to whether $E_1$ occurred. So we require that the chance of $E_1$ given $E_2$ is the same as the chance of $E_1$: we require that $P(E_1|E_2) = P(E_1)$. By the definition of conditional chance, this means that $P(E_1 \cap E_2) = P(E_1)P(E_2)$. More generally, if the $n$ outcomes $E_1, E_2, \ldots, E_n$ are independent, we must have,

$$P(E_i \cap E_j) = P(E_i)P(E_j) \quad \text{for} \quad i \neq j,$$
$$P(E_i \cap E_j \cap E_k) = P(E_i)P(E_j)P(E_k) \quad \text{for} \quad i \neq j \neq k \neq i$$

and so on until

$$P(E_1 \cap E_2 \cap \ldots \cap E_n) = P(E_1)P(E_2) \ldots P(E_n).$$

*Independent trials*

Yet another idea needs precise definition. It is often believed that tosses of a coin are made under identical conditions, so that the chance properties of the system remain the same from trial to trial. This conception differs from the one just studied: we were able to define independence of outcomes on trials of some kind, but now we want to explain independence of trials themselves. Fortunately the new notion can be defined in terms of the preceding one.

If trials themselves are to be independent of each other, then, for instance, the outcome of one trial should have no influence on the outcome of the next. So we have to consider a pair of trials together. Pairs of trials naturally form a single *compound* trial. If a compound trial consists of two simple trials, its outcome has two parts, (i) the outcome of the first simple trial, and (ii) the outcome of the second simple trial. If the first simple trial has outcome $E_1$, the compound certainly has outcome $\langle E_1, \Omega \rangle$; if the second trial has outcome $E_2$, the compound certainly has outcome $\langle \Omega, E_2 \rangle$. I shall call these the first and second components of the outcome $\langle E_1, E_2 \rangle$. If the simple trials are independent, then the components of our compound trial must be independent in the sense of the preceding section. So we shall be able to define one aspect of independence in terms of our earlier work.

There is another aspect. If trials are independent, the long run frequency of any outcome $E$ among trials in general should not differ from its frequency among, say, odd-numbered trials. We shall have to build this feature into our definition. Von Mises has given a colourful description of independence. Suppose the chance of some event $E$ on trials of kind $K$ is $p$, and that whenever a gambler bets on trials of kind $K$, he must either bet on $E$ at odds of $p:1-p$, or against $E$ at the reverse odds. Then, says von Mises, if trials are independent no long run successful gambling system should be possible. This is closely connected with the preceding remarks. Any gambling system must be a combination of two kinds of system: *static* and *dynamic*. In a static system you decide ahead of time which tosses to bet on, and how to bet at each toss. In a dynamic system you lay your bets in accord with the outcomes of preceding tosses. For instance, if the outcome of one trial influences the outcome of the next, a successful dynamic system may be possible. We shall preclude this in the first part of our definition of independence. Secondly, if the long run frequency of heads in odd-numbered tosses differs from that in tosses in general, a static system, based on betting the odd tosses, will certainly be successful. We aim to exclude this with the second condition.

The details are as follows. Let there be given trials of kind $K$. By a compound trial I shall here mean a sequence of $n$ consecutive trials of kind $K$. If the $n$ trials have results $E_1, E_2, ..., E_n$, the compound trial has as result the ordered set $\langle E_1, E_2, ..., E_n \rangle$. A compound kind of trial is one in which any particular trial is as just described, and in which the next trial of the same kind is based in the same way on the next $n$ consecutive trials of kind $K$. Evidently outcomes of a compound kind of trial are composed of outcomes of the $n$ trials of kind $K$. The first component of an outcome on the compound trial will be $\langle E_1, \Omega, ..., \Omega \rangle$; the second component, $\langle \Omega, E_2, \Omega, ..., \Omega \rangle$, and the last $\langle \Omega, ..., \Omega, E_n \rangle$. For convenience represent the $k$'th component by $\langle E \rangle^k$. Thus a compound trial has outcome $\langle E \rangle^k$ if the $k$'th simple trial making up the compound has outcome $E$.

Finally our definition: Trials of kind $K$ are *independent* if

(1) the components of any compound kind of trial based on trials of kind $K$ are independent (in the sense of the preceding section); and,

(2) for every outcome $E$ of a trial of kind $K$, and for every compound kind of trial based on trials of kind $K$, the chance of $\langle E \rangle^k$ on the compound kind of trial equals the chance of $E$ on trials of kind $K$.

This definition is not as strict as some which might be given, but in practice it suffices for all the usual mathematics. The main use of a definition of independent trials is in computing the chance of some outcome of a sequence of $n$ (not necessarily consecutive) trials

INDEPENDENT TRIALS

of some kind. This can always be achieved from our definition. Moreover it is fairly easy to provide infinite sequences which can serve as models of results of independent trials. But it will be noted that our definition of independence lays no condition on 'the long run'. Ultimately we have defined independence of trials in terms of long run frequency on trials of some kind; whatever ambiguity lies in 'long run frequency' is retained in our definition of independence.

*The binomial distribution*

It is instructive to run through a well-known derivation in order to show the power of our definitions. What is the chance of getting $k$ heads in $n$ independent tosses with a coin? More exactly, if the chance of heads on independent trials is $p$, what is the chance of getting $k$ heads on a compound trial consisting of $n$ consecutive single trials?

I shall develop the answer in some detail. First observe that one possible result of the compound trial is $\langle H, H, H, ..., T, T, T\rangle$, where $k$ heads are followed by $n-k$ tails. The components of this outcome are $\langle H \rangle^1, \langle H \rangle^2, ..., \langle T \rangle^{n-1}, \langle T \rangle^n$. According to our second condition on independent trials, the chances of the first $k$ components must each be $p$. According to the same condition, plus Kolmogoroff's axioms, the chances of the last $n-k$ components must be $1-p$. The chance of the whole result $\langle H, H, ..., T, T\rangle$ must, by our first condition on independent trials, equal the product of the chances of the components, namely $p^k(1-p)^{n-k}$. Likewise for any result with just $k$ occurrences of heads and $n-k$ of tails. By the Kolmogoroff axioms, the chance of getting $k$ heads in $n$ tosses must be the sum of the chances of all such results. There will be as many such possible results as there are ways of picking $k$ things out of $n$, namely $n!/k!(n-k)!$. Hence the chance of getting $k$ heads in $n$ tosses must be

$$\frac{n!}{k!(n-k)!}p^k(1-p)^{n-k}.$$

This formula, discovered by Newton, was put to this purpose by James Bernouilli. Notice that it provides the distribution of chances on compound trials consisting of $n$ simple trials of kind $K$, for we may, from the formula, compute the chance of each possible result of the compound trial. This formula, called the binomial formula, thus leads us to call the distribution the *binomial distribution*. I have worked it out for the case of $n$ consecutive tosses, but it is obviously applicable to any case of $n$ different tosses of the coin.

## Kinds of trial

The property up for definition—chance, or frequency in the long run—is now seen to be a property of a chance set-up; more expressly, the chance of a designated outcome of a stated kind is a property of a specified chance set-up. The 'kind of trial' must never be forgotten, even if it is convenient sometimes to omit it by way of ellipsis.

One false objection could result from ignoring the kind of trial. Suppose that on a particular toss of a coin, say $m$, the coin falls heads. Then it may be argued that, on this toss, the true chance of heads must have been 1. After all, the argument would continue, the coin was caused to fall as it did even if the determining factors are unknown, so the only true record of the propensities in this case is $P(H) = 1$. Since this opinion holds for any toss whatsoever, the chance of heads from any specified toss must be either 0 or 1. Perhaps in a set-up studying particular photons, whose behaviour is, on current theory, not fully determined, it may make sense to talk of chances between 0 and 1, but not with coins.

Several superficial replies are possible, but in the present theory of chance, the crux of the matter is this: on what kind of trial is it alleged that $P(H) = 1$? It simply does not make sense to talk about the chance of heads unless the kind of trial be specified. I guess that the kind of trial most likely to be cited in the above argument is that kind which includes only tosses in which exactly the same causal influences are at work, in exactly the same proportions, as in the toss $m$. However one imagines this ought to be analysed, let us call this kind of trial $M$. Perhaps only one trial of this kind is possible, namely, the trial $m$. But I have no desire to exclude such trials from the present theory. Say indeed that on trials of kind $M$, $P(H) = 1$. This is simply a corollary of determinism, and far be it from a theory of chance to preclude that august philosophical ogre. But there are trials other than those of kind $M$. Say that a trial of kind $K$ consists of tossing this coin from this device and noting the result. Then it is perfectly consistent to say (i) on toss $m$, the coin fell $H$, (ii) on trials of kind $M$, $P(H) = 1$, and (iii) on trials of kind $K$, $P(H) = 3/4$. There is nothing unusual in regarding one event under several aspects. Observe that $P(H)$ on trials of kind $M$, and $P(H)$ on trials of kind $K$, may both be properties of the chance set-up; perhaps the latter property is just the consequence of a collection of properties of the former sort, but anyway the two are consistent, and the current doctrine of chances is not impugned.

## Populations

I follow Venn and von Mises, intending to explain frequency in the long run by means of chance set-ups. Others, with a very similar aim, and a similar property in mind, take properties of 'populations' as their prime concern; in particular, the frequency with which some characteristic is found in a possibly infinite population. It can be argued that the great statisticians—Laplace, Gauss, Fisher, the Pearsons, even Neyman and Wald—conceived frequency in this way. It is a natural way if knowledge about frequencies is not an end in itself, but only a tool in the reduction of complex data; as when one tests the efficacy of a technique of immunization by comparing the frequency in the long run of infection amongst treated and untreated patients. Fisher said that the solution of such problems 'is accomplished by constructing a hypothetical infinite population, of which the actual data are regarded as constituting a random sample',† and goes on to discuss the laws of distribution of members of such an hypothetical population, comparable to the laws of distribution of a chance set-up.

For all the prestige which great thinkers have given the idea of infinite populations, it is hard for an onlooker to conceive frequency other than in terms of something like a chance set-up. A procedure for sampling a population certainly involves a chance set-up. But only excessive metaphor makes outcomes of every chance set-up into samples from an hypothetical population. One hopes our logic need not explicitly admit an hypothetical infinite population of tosses with this coin, of which my last ten tosses form a sample. Chance set-ups at least seem a natural and general introduction to the study of frequency. Especially since they lend

† R. A. Fisher, *Statistical Theory of Estimation* (Calcutta, 1938), p. 1.

themselves naturally to the idea of a chance process; to describe a chance process in terms of samplings from populations, you probably need an hypothetical infinite array of hypothetical infinite populations. Chimaeras are bad enough, but a chimaera riding on the back of a unicorn cannot be tolerated.

Fisher's remark recalls that when frequency is taken as a characteristic of populations investigated by sampling, the sampling is generally supposed to be random. But randomness in this context can only be explained by frequency in the long run. So there is some danger of a silly circle. The danger is skirted by a study of chance set-ups. One need never mention randomness until one needs it, namely when inferring characteristics of some real population (say the 9200 English vegetarians) from established facts about the distribution of chances of outcomes obtained by sampling the real population. By the time it is needed, randomness can be effectively defined in terms of chance set-ups. It will be done in ch. VIII below.

None of these remarks is intended to deny the use of populations in the statistical imagination. It questions their value in logic. The objection is not only to the infinity of Fisher's hypothetical populations. Braithwaite has shown how to avoid that. But the chance set-up is the natural background for thinking of frequency, and I shall try to prove it is the simplest primitive notion for statistical inference.

CHAPTER III

# SUPPORT

Kolmogoroff's axioms, or some equivalent, are essential for a deep investigation of our subject, but do not declare when, on the basis of an experiment, one may infer the exact figure for the frequency in the long run, nor the distribution of chances on trials of some sort. They do not even determine which of a pair of hypotheses about a distribution is better supported by given data. Some other postulates are needed.

In what follows a *statistical hypothesis* shall be an hypothesis about the distribution of outcomes from trials of kind $K$ on some set-up $X$. Our problems are various: given some data, which of several statistical hypotheses are best supported? When is a statistical hypothesis established, and when refuted? When is it reasonable to reject an hypothesis? We should also know what to expect from a chance set-up, if a statistical hypothesis is true.

Each question is important, and the correct answers, in so far as they are known, already have a host of applications. But the first question is crucial to the explication of chance. If no one can tell in the light of experimental data which of several hypotheses is best supported, statistical hypotheses are not empirical at all, and chance is no physical property. So our first task is to analyse the support of statistical hypotheses by experimental data.

It is common enough in daily life to say that one hypothesis is better supported than another. A man will claim that his version of quantum theory is better supported than a rival's, or that while there is little support in law for one person's title to his land, there is a good deal to be said for someone else's claim. Since analysis must begin somewhere, I shall not say much about the general notion of support. But I shall give a set of axioms for comparing support for different hypotheses on different data, and, very much later, axioms concerning the measurement of support-by-data. It is to be hoped that the common understanding of the English word 'support' plus a set of axioms declaring the principles actually to be used, will suffice as a tool for analysis.

Instead of the notion of support I might equally well begin with the 'probability' which data provide for an hypothesis. This is a usual custom. But 'probability' has been used for so many things that I avoid it, just as I avoided it for long run frequency. I do not want to invite misunderstanding, and so I shall speak of support for hypotheses by data. It is, of course, no accident that our axioms for comparative support are taken from well-established probability axioms.

A few special facts about support are worth noting. If one hypothesis is better supported than another, it would usually be, I believe, right to call it the more reasonable. But of course it need not be reasonable positively to believe the best supported hypothesis, nor the most reasonable one. Nor need it be reasonable to act as if one knew the best supported hypothesis were true.

Since these obvious facts seem to have been denied a trifling story may make them yet more plain. A young man is informed that a scented letter awaits him. He knows it must be from Annabelle or Belinda. The former, whom he admires, writes but seldom and erratically, while the latter, whom he dislikes, is his constant correspondent. Knowing nothing else, and having no especial reason to expect a letter from either, his best guess is that the letter is from Belinda. That she wrote is the most reasonable hypothesis, and the one best supported by the frugal evidence. But it need not be reasonable for the man positively to believe the letter is Belinda's. He should believe only that this is very probable. Nor need it be reasonable for him to act as if the letter were hers, or as if he believed this were true. For to act that way—is to tear up the letter unopened. A most unreasonable procedure if there is still a chance that the letter is from his true love.

Even if these points are evident a related confusion is more tempting. Suppose one wishes to know the chance of heads from a set-up. There are two quite distinct questions: (1) Which hypothesis about the true value is best supported by current data? (2) In the light of the data, which is the best estimate of the true chance? The questions differ because in at least one way of understanding the matter, an estimate is good or bad according to the purpose to which it must be put. A general's best estimate of the number of his troops will very often be an underestimate; his opponent's best estimate of the same number could be an overestimate. Unless one specifies or assumes a purpose, there may be little point in asking after the best estimate. But one may perfectly well determine which hypothesis is best supported without weighing ends in a similar matter. Support is a concept independent of utility.

Even in pure science where the sole aim is supposed to be truth, this distinction is as important as ever. Speaking very intuitively for a moment, an estimate is good if it is very probable that the true value is near it. But an hypothesis is not best supported according as it is probable or not that the truth lies near the hypothesis. To take a crude but instructive example, suppose there are six hypotheses about the value of $A$, namely $A = 0.90$ or $0.12$ or $0.11$ or $0.10$ or $0.09$ or $0.08$. Suppose that the last five are equally probable—in any sense you care to give 'probable'—and that the first is slightly more probable. Then, if one may infer that the most probable hypothesis is best supported, $A = 0.9$ is best supported. But it is much more probable that $A$ is near to $0.1$, and so $0.1$ may be the best estimate. Of course every term, including 'near', must be made precise before this is a clear-cut example, but it suffices here to establish a difference between 'best-supported' and 'best estimate'.

I insist on these distinctions because it is now customary to include many or even all statistical questions under the general head of *decision theory*. This is the theory of making decisions when you are not completely certain of their outcome. The general form of problems currently treated is as follows: *Given* (1) a set of possible hypotheses about the state of the world, (2) some experimental data, (3) some possible decisions, (4) the value of the consequences of each of these decisions under each possible state of the world; *determine* the best decision. The general solution of the problem consists of a function mapping a unique decision on to each possible experimental result. Modern methods of treating the problem stem almost entirely from a powerful analysis due to Wald.

Evidently each of the four numbered items is relevant to a decision. Although there may be intelligent decisions even when there are virtually no experimental data, there can be none unless one has some inkling of the value of the consequences of one's decisions. Hence the weighing of utilities—losses, gains, value, profits, risks, to take the commercial terms—is an essential prerequisite to any application of decision theory.

At least part of statistics can be plausibly assimilated to decision theory. In recent years this fact has dominated much research, and has led to wonderful discoveries. But it does not follow that decision theory is a fitting vehicle for the philosophical elucidation of chance. In particular, the concept of best supported hypothesis does seem independent of utilities, so it should be possible to refine this concept without decision theory. Not only should it be possible, but several facts suggest it is necessary.

First, much current decision theory is about what to do, assuming that there is an empirically significant physical property of chance. This assumption is of course correct. But our philosophical task is to see why and how it is correct. A theory which assumes it is correct is not likely to show why it is correct.

Secondly, decision theory requires some conception of utilities as a prerequisite. Quite aside from the fact that support, at least, does not involve utilities, it cannot be denied that utilities are at least as obscure as chance; probably a good deal more so. Their definition is still the centre of controversy. It would be sad if we had to import another disagreement into philosophical discussions of chance. It may even be dangerous. According to one school, utilities can only be defined, in full, in terms of chance or something similar. In another school, utilities are based on the readiness of a person to make various gambles in which he knows the chances of success. So chance is probably the prior notion.

Less important but still relevant, most current decision theory advises decisions on the basis of, in part, the *expectation* of loss. This is by no means an essential feature of decision theory, but is characteristic of most current work. But the expectation is defined by a weighted sum or integral of chances. So, from a philosophical point of view, it is better first to get at chance, then decisions based on it.

Despite all these reasons, some readers will find it repugnant that we dissociate ourselves from an important part of current statistical thought. Perhaps the theory of support can be brought, formally, within the structure of decision theory, perhaps, they may feel, all that is needed is assigning a fake utility $+1$ to the decision that an hypothesis is best supported, when it is in fact true, and $-1$ to the same decision when the hypothesis is false. But I think this dubious in detail; even if not, it should be unnecessary. In philosophy, every unnecessary notion increases the chance of error many times.

Or some philosophers will be impressed by the recently canvassed idea of epistemic utility which is an attempt at measuring what the value of a proposition would be, if it were known; the epistemic utility of a proposition is in part a function of what is already known.† Ingenious as is this idea, it is still so far from a statement which is both plausible and precise that it cannot at present give any foundation to a study of chance. Nor is it evident that anyone ever decides anything on the basis of the epistemic utility of an unknown proposition unless, indeed, it is the decision to find out if the proposition is true.

Two hitherto unmentioned facts may incline some people to include the theory of support in decision theory. First, to conclude that an hypothesis is best supported is, apparently, to decide that the hypothesis in question is best supported. Hence it is a decision like any other. But this inference is fallacious. Deciding that something is the case differs from deciding to do something. Sometimes they may be parallel, especially in expressing intentions. Deciding that I shall take a walk is generally the same as deciding to take a walk. But sensible deciding that something is the case need not involve weighing the utility of this decision, while sensible deciding to do something does involve weighing consequences. Hence deciding to do something falls squarely in the province of decision theory, but deciding that something is the case does not.

Another consideration is less important. Several modern writers imply that deciding something is the case is deciding to assert something. Deciding to assert is deciding to do, and so falls under decision theory. But in fact, deciding that something is the case is not deciding to assert it. One may of course decide that something is the case and then assert it, or even blurt it out, but that does not mean, necessarily, that one decided to assert it. Most people do most things without deciding to do them. On the other hand, one may perfectly well decide that something is the case, and at

† C. G. Hempel, 'Deductive nomological *vs.* statistical explanation', *Minnesota Studies in the Philosophy of Science*, III (1962), 154.

the same time decide not to assert it. This may happen in the purest science; it is the plot of Dürrenmatt's chilling play, *The Physicists*.

## The logic of support

The logic of support has been studied under various names by a number of writers. Koopman called it the logic of intuitive probability; Carnap, of confirmation. Support seems the most general title. A piece of evidence may well give some support to a proposition without confirming it. But the name does not matter much. We could perfectly well have stipulated that 'probability' should mean support.

I shall use only the logic of comparative support, concerned with assertions that one proposition is better or worse supported by one piece of evidence, than another proposition is by other or the same evidence. In this respect the following work will differ from any previous study.

The principles of comparative support have been set out by Koopman; the system of logic which he favours will be called Koopman's logic of support.† It will be indispensable to the rest of the work. To state the logic exactly we first assume a *Boolean algebra* of propositions, that is, a set of propositions closed under the operations of alternation ('or') and negation ('not'). This means that if $h$ and $i$ are propositions in the set, so is $h$-or-$i$, and also not-$h$. As is customary in logical work, we represent 'or' by v, and 'not' by $\sim$. In the case where there are infinitely many propositions under consideration, we shall insist that our algebra of propositions is closed under countable alternation. If $h_1, h_2, \ldots,$ is an infinite list of propositions, then the countable alternation of these propositions is that proposition which is true if and only if some proposition in the list is true. A set of propositions closed under negation and countable alternation is called a *Boolean sigma-algebra*.

Given such an algebra of propositions, the following notation is useful. $h/d \leq i/e$ shall signify that $e$ supports $i$ at least as well as $d$ supports $h$. The only theses from Koopman's logic which we shall need are:

† B. O. Koopman, 'The axioms and algebra of intuitive probability', *Annals of Mathematics*, XLI (1940), 269–92,

(1) *Implication.* If $h$ implies $i$, then $h/e \leq i/e$. Here and elsewhere, 'implies' shall be used to mean logically implies, so that $h$ implies $i$ only if $i$ is a logical consequence of $h$.
(2) *Conjunction.* If $e$ implies $i$, then $h/e \leq h \& i/e$.
(3) *Transitivity.* If $h/e \leq i/d$ and $i/d \leq j/c$, then $h/e \leq j/c$.
(4) *Identity.* $h/e \leq i/i$.

Typical consequences of these theses are:
$$h/e \leq h/e.$$
If $e$ implies $h$ and $d$ implies $i$, and $j \& h/e \leq k \& i/d$, then $j/e \leq k/d$.

*Comparative and absolute support*

It is very important to distinguish the comparative logic of support from the far stronger logic of 'absolute' quantitative support. In that logic it is supposed that there are quantitative measures of the degree to which an hypothesis $h$ is supported by evidence $e$. The axioms usually proposed for this notion are the very same ones we have accepted for frequency. In our study, $P(E)$ is interpreted as the chance of $E$; and $P(E|F)$ as the chance of $E$, on condition that $F$ occurs. But many other interpretations are possible. It has been proposed that $P(E|F)$ be construed as the degree to which an hypothesis $E$ is supported by an evidential proposition $F$. This interpretation has nothing to do with any physical property.

There are plenty of well known riddles about this new interpretation. I am not sure of the solution to any of them. But for the present they can very largely be ignored. The difficulties do not show up in the logic of comparative support. But there is one possible defect in Koopman's logic which had better be recorded. His logic must be a trifle too strong, for it includes the principle of *anti-symmetry*:

If $h/e \leq i/d$, then $\sim i/d \leq \sim h/e$.

Doubt is cast on the principle if we put $e = d =$ 'a box contains 40 black balls and 60 green ones; a ball will shortly be drawn from the box'. Let $i =$ 'the next ball to be drawn from the box will be black', and let $j =$ 'the next ball to be drawn from the box will be green'. I take it that $e$ supports both $i$ and $j$; without $e$ or some other piece of evidence, there would be no support for either of those propositions, but given $e$, there is some support. Now $i$ and $j$ are contraries, so $j$ implies $\sim i$. By the thesis of implication quoted earlier, it follows that $e$ supports $\sim i$ at least as well as $j$. Hence $e$ furnishes support both for $i$ and $\sim i$.

Finally take $h = $ 'there are fish in the Dead Sea'. I take it that $e$ furnishes no support whatever for this proposition or its negation. Hence I conclude that

$$h/e < i/e, \quad \text{while also} \quad \sim h/e < \sim i/e,$$

contrary to anti-symmetry.

This seems to me a forceful argument, although like many other counter-examples to principles of support, it can be subject to various kinds of hair-splitting. Perhaps it shows that Koopman's system should be restricted to pairs of propositions, $h$ and $e$, such that $e$ is relevant to $h$ or $\sim h$. This restriction will be tacitly granted throughout the rest of this book. But it would be pleasant to discover a logic of support slightly weaker than Koopman's, and which is not open to this kind of objection.

*The rôle of logic*

Koopman's logic of support will underlie all our future work, so it may be worth giving a brief account of the rôle of logic in an investigation like ours. It will also be useful to try to dispel the idea, popular among non-logicians, that logic is a kind of fixed immutable science. In fact, like every other human endeavour, it is just a patchwork quilt whose patches do not meet very well, and which are continually being torn up and restitched.

Uses of logic can be described very formally, but an informal approach is more easily understood. First the idea of a *theory*. Etymologically, a theory is a collection of speculations. Logicians have revived this usage, and mean by a theory the set of truths about some subject. Thus arithmetic is the theory of numbers, and geology, I suppose, is the theory of rocks. Statistics is the theory of chance.

The logical analysis of a theory will take for granted some *underlying logic* and to this add a further set of *postulates* about the theory. No one has ever exactly defined the idea of an underlying logic. Perhaps any set of principles, believed to be consistent, and

## THE RÔLE OF LOGIC

thought of as having a logical character, might serve as an underlying logic in some investigation. Thus in the analysis of arithmetic, logicians typically take for granted (*a*) a logic of prepositions, including principles and rules bearing on terms like 'not' and 'or' and 'implies'; (*b*) a logic of functions, including principles bearing on terms like 'all'; (*c*) a logic of sets, including principles bearing on terms like 'is a member of'. Different schools of thought use different logics, though there is a large measure of agreement on (*a*), some on (*b*), and a kind of cheerful concordat about (*c*). Of course anyone studying the theory of sets would take (*a*) and (*b*) as his underlying logic, and try to improve the (*c*) for future workers, while someone studying a very special field might supplement (*a*)–(*c*) by whatever he needs.

The *closure* of a set of postulates is the set of propositions derivable from the postulates in accord with the underlying logic. Thus the underlying logic determines the closure. A member of the closure will be called *non-logical* if it does not follow from the underlying logic alone, but requires the special postulates. Then a theory is a *model* for a set of postulates if every non-logical member of the closure is a truth of the theory. The set of postulates is said to be *complete*, or to characterize the theory completely, if the theory is a model for the postulates, and if every truth of the theory is a member of the closure.

Evidently the logical analysis of a theory aims at completeness. But here we must distinguish two situations. It may happen that the theory itself is precisely characterized independently of our logical investigation. That is, it is in some sense an entirely definite question whether or not an arbitrary proposition is a member of the theory. Then it is presumably a clear-cut mathematical problem to discover whether or not the set of postulates is complete. But often the whole point of the logical investigation is to characterize the theory for the first time. Then it might no longer be a definite mathematical question, whether or not a set of postulates completely characterizes the theory. For we have, on the one hand, a theory which is not precisely defined; on the other hand, something entirely precise, namely an underlying logic and a set of postulates. There is no obvious sense in saying, in every such case, that the precise thing is exactly equivalent to the imprecise thing. However some question about completeness is still significant. To say the theory is imprecisely defined is not to say we know nothing about it. If anyone is seriously trying to analyse some theory, he presumably thinks he knows some truths of the theory. Thus he is led to a sort of inductive procedure. In the words of *Principia Mathematica*, he must try to show that everything which is probably true in the theory does follow from his postulates, and that nothing probably false does follow. Perhaps that is all he can do; it is a situation which some kinds of mathematician find unpalatable, but they are free to work at other things. Students with a philosophical bent, or with a genuine interest in the theory, will turn themselves to the 'inductive' question, repeatedly testing the postulates in terms of their consequences.

Moreover, two kinds of purely mathematical result still remain. First, a set of postulates might be so strong that adding any underivable proposition—or any underivable proposition of a certain sort—would simply lead to contradiction. Then, if the postulates are truths of the theory, they must completely characterize at least some aspect of the theory. We shall have one statistical instance of this towards the end of the essay. Secondly, *incompleteness* might be demonstrable. Even if a theory is not precisely defined, it may still be possible to know that the proposition $A$ is a truth of the theory, and yet is not in the closure of the postulates. The most celebrated of all logical discoveries is of this sort. Gödel proved that *any* consistent set of postulates adequate for a central branch of arithmetic must be incomplete in just the above sense. No such result is to be expected at the present stage of foundational inquiry into statistics.

Rigorous mathematicians may shy away from the 'inductive' testing of a set of postulates in terms of their consequences, but philosophers might find it equally distasteful. For concreteness, let the theory under consideration be statistics. I glibly say we know some truths of this theory. What is the nature of these truths? Necessary or contingent, analytic or synthetic, *a priori* or *a posteriori*? I seem to write as if there were an abstract body of statistical truths, existing independent of what is known, and which comprise our theory. How could I be sure of this? In reply I wish simply to ignore all such questions. Some simple propositions about chances are generally agreed in statistics, and may for the moment be regarded as probably true. In saying this, and in speaking of the theory, I do not commit myself to any view on the nature of truth. I could hold an entirely stupid form of subjectivism, and say that what is true is what people continue to agree on, or I could hold an entirely stupid form of objectivism, imagining eternal and inhuman propositions labelled true or false. Whatever be right about truth, it is possible to know of some statistical propositions which are probably true. Indeed, if there were none such there would be no statistics to investigate.

I do not wish to suggest that questions about truth are unimportant. I mean to imply only that they are hard. If I could say what it is for something to be true I would say it, but I cannot. Nor can anyone else, yet. Success in that enterprise is no prerequisite for scientific progress. Nor should we have the picture of building, step by step, on immutable pieces of knowledge. Rather, gladly admit that some things about statistics are agreed on, and use these as a basis for testing postulates. In the end, men may reverse the business and test what is now agreed against some compelling body of postulates. But that is a prospect for the future, when our science is stable. Recall Neurath's well-known simile: scientists are like sailors trying to rebuild a ship on the open sea. In the end every plank may be changed, but at any stage there are some planks we leave alone. At our stage in our investigations, we shall not tamper much with the most widely agreed facts about statistics, though indeed we shall be ruthless with many theories which, however widely held, are intended to explain those facts.

Finally I may return to particulars. Statistics, being a relatively specialized science, much more specialized than arithmetic, does call for an extensive underlying logic. We shall take any standard logic currently effective for mathematics, and to this add Koopman's logic of support. The resulting underlying logic will be called, for short, Koopman's logic; we simply take for granted what is standard everywhere in mathematics. To our underlying logic we shall add postulates about chance. Three have been added already, in the form of Kolmogoroff's axioms. Now we search for some more.

The idea of using something like a logic of support as a basis for statistics is very old. It is most successfully and self-consciously applied in Jeffreys' *Theory of Probability*. But his underlying logic is stronger than Koopman's, and his special postulates have seldom seemed correct. They will be discussed much later. For the present we remain with Koopman's logic and try to find the postulates needed both for the definition of chance and for the foundations of statistics.

CHAPTER IV

# THE LONG RUN

The logic for support will serve as the underlying logic for our definition of chance. So now we seek connexions between support and long run frequency. There is one connexion which has usually seemed too obvious to be worth the stating, but this whole chapter is devoted to it. It serves to bring out the rôle, or lack of rôle, of the concept of the 'long run' in statistical inference and in reasoning about chances. Statisticians have usually been interested in new discoveries, but a philosophical inquiry into foundations has to begin with what everyone else takes for granted.

The suitor of the last chapter guessed who wrote a letter; he knew only that one correspondent writes more often than another. It seems part of our very understanding of frequency in the long run, that if $A$ happens more often than $B$, and one or the other must have happened, then the best guess is $A$. This sounds correct no matter what one means by the long run. It does not follow from Kolmogoroff's axioms. Should it be derived from another postulate? Must it be postulated on its own? Or is it just false?

The questions are academic. Seldom are frequencies our only data. The world is too complicated and men know too much. Only in the imagination do frequencies serve as the sole basis of action. But it is probable that laws governing imaginary cases operate in life.

Suppose an urn to contain many balls. Most are black; a few white. Experiments show that when a blindfold person draws a ball at random, replaces it, and shakes the urn before redrawing, black appears far more often than white. So suppose this is a chance set-up in which the chance of black greatly exceeds that of white, and in which the result of any draw is independent of any other. If you are to guess about the result of the next draw, or any other, then, on this slender information, your best guess is 'black'. Why?

Black is best only if you want to guess correctly. I shall take this for granted in the sequel, but perhaps it is not a necessary assumption. Instead of guessing one may speak more formally. The hypothesis of a black ball being drawn next is much better supported, by the slim data which have been stated, than is the hypothesis of white. Briefly, 'black' is better supported than 'white'. Again, I ask, why?

One answer is all too readily forthcoming. I shall call it the *long run justification*: if you are to guess on a long sequence of like draws, and want to make as small a number of wrong guesses as possible, you should guess 'black' every time. So 'black' is the best guess about any particular draw, regardless of how many draws are actually to be made, and 'black' is the best supported hypothesis. 'If we are asked whether or not the side 1 of a die will appear in a throw, it is wiser to decide for "not-1" because if the experiment is continued, in the long run we will have a greater number of successes.'† This answer is given by men in all walks of life, and is accepted in much statistics and philosophy. It is wrong.

First a caution about the 'long sequence'. Whatever one means by frequency in the long run, I take it that no infinite sequence, or run, is mentioned in the long run justification. There is no clear sense to a man's guessing infinitely often; even if there were, success in that misty realm would be no argument for any more human pattern of behaviour. I take it that the long sequence referred to is a long run of 200 or 2000 or even 2,000,000 guesses. Nothing specific, just long in some human scale of length.

The long run justification defends what is essentially a weak rule about guessing. For all $A$ and $B$, and any unique specified thing $T$, if experiments show that $S$'s are either $A$ or $B$, never both, and more frequently $A$ than $B$; if it is known that whether any particular $S$ is $A$ is independent of other $S$'s, and if it is known that $T$ is an $S$, then the hypothesis that $T$ is $A$ is better supported by this data than is the hypothesis that $T$ is $B$.

This will be called the *rule for the unique case*, since it can be used for guessing on any uniquely specified thing, such as the next draw from an urn, or a particular scented letter. No long run of guesses is required. Perhaps the rule implies that if you want to guess correctly, and must guess on $T$ but not necessarily on anything else, then your best guess is $A$. If not, another version of the rule will state this.

† H. Reichenbach, *Experience and Prediction* (Chicago, 1938), p. 310.

Parallel to the unique case rule is what I shall call the *rule for the long run*: under exactly the same conditions, if one is to guess on each of a long sequence of $S$'s, and wants to be right as often as possible, the best policy is to guess $A$ every time. Once again, I suppose the sequence is long in some human scale. There are plenty of other possible rules for long run behaviour, but this one is commonly singled out as best, and so is dignified by our name, *the* long run rule.

The long run justification may be rephrased as arguing that the long run rule is obviously correct, implies that $A$ is the best guess in a unique case, and implies the unique case rule. In most of what follows I won't question the truth of either rule. I do question the alleged justification, and contend that it is false. Not only does the long run rule not formally imply the unique case rule, but also there is no valid way of inferring the one from the other. Now I am not much interested in the long run rule, but want to ask why it is 'obviously correct', in order to discover whether the reasons for accepting it, if any, can be transferred to the unique case rule.

## Goodman's riddles

Before proceeding to justification we must, in aside, mention *Goodman's riddles*. Let 'blight' mean 'black until the end of 1984 and white thereafter'; let 'wack' mean 'white until the end of 1984 and black thereafter'. Goodman argues that every shred of evidence which supports the claim that most balls in an urn are black, and that black is drawn correspondingly more often, equally supports the claim that most are blight, and that blight is drawn more often than wack. So it may seem that the long run rule is either indifferent between the policy of guessing black, and the policy of guessing blight (i.e. guess black till the end of 1984, and white thereafter) or else that it recommends one guess black every time, and also that one guess blight every time. Either alternative is absurd.

Goodman puts his riddles so cogently that I hardly need emphasize the dilemma.† They combine precision of statement, generality of application, and difficulty of solution to a degree greater than any other philosophic problem broached in this

† N. Goodman, *Fact, Fiction and Forecast* (Cambridge, Mass., 1955).

century. Goodman thinks we must divide predicates into two classes, 'projectible' (those which can be 'projected' into the future) and the rest (those, like wack, which we do not want to 'project'). So perhaps the long run rule should insist on $A$ and $B$ being projectible predicates. But I shall postpone the difficulty, and put the onus on 'shown by experiment or experience' as it appears in the rubric of the rule. Currently practicable experiments can show, and justly show, that black, as opposed to blight, is drawn more often in the long run. Exactly why they show this is very hard to state. That is why Goodman's riddles are so pressing. But the fact cannot be denied that black as opposed to blight can be now established as the most frequently drawn ball. Hence the long run rule can instruct us unequivocally to guess black every time. Of course those who would not put the onus on the undefined 'shown by experiment' can add to the long run rule the condition that $A$ and $B$ are 'projectible'.

*Arguments based on success*

First, some specious justifications of the long run rule. Some may prefer it because they think it's certain to be more successful in the long run than any other rule. They think someone using it will make fewer wrong guesses than someone using any other rule.

But it is certain that, on any particular sequence with some $B$'s in it, another rule of guessing would make fewer errors than the long run rule. If the first $S$ were $A$, the next $B$, the next $A$, and so on in some order, this better rule would begin, 'guess $A$, then $B$, then $A$...'.

For very many sequences a rule like '$AAB$'—alternate two guesses of $A$ with one of $B$—might be more successful than '$AAA$', or guess $A$ every time. Indeed one cannot tell beforehand whether '$AAB$' or '$AAA$' will be better. I am here remarking only that it is not certain that '$AAA$' will fare better. It is certain that some other rule will fare better than it.

It will be retorted that, despite occasional superiority of '$AAB$', the long run rule will be more successful in the long run of long sequences. I object on three counts. First, this is by no means certain. Why should not '$AAB$' be just a little bit better than '$AAA$' on some long run of long runs? Secondly, even if it were certain that using '$AAA$' on every sequence would be better than using '*AAB*' on every sequence, why should not one use '*AAA*' on some sequences, '*AAB*' on others, and come out better than if he had used '*AAA*' every time? It is not certain that he would not. Thirdly, even if one should use '*AAA*' on every long sequence in a long sequence of sequences, does it follow that '*AAA*' is the right rule for just one long sequence? It follows, apparently, only if some form of the unique case rule be assumed. But if it has to be assumed, the long run rule cannot be used to justify the unique case rule.

*A minimax argument*

There is another argument for the long run rule which seems to me dubious, but which I will defend as best I can. If my defence is invalid, so much the worse for the long run justification, for I know of no other argument for the long run rule which has any merit.

The long run rule is not certain to make fewer errors than any rival. But the rule is to be applied only when experiments show certain facts to hold. These facts may be thought to guarantee, in a certain way, that no more than a certain proportion of guesses made by this rule will be wrong. They do not guarantee so small a proportion for any rival rule. Very roughly, the long run rule cannot be wrong more than half the time, while '*AAB*' might, for all that is known, err on 2/3 of the possible guesses.

It is very tricky to show this. To do it at all one must assume for the sake of the argument some statistical theory of testing hypotheses. But I am prepared to grant so much theory to those who admire the long run rule, in order that they have as good a case as possible. Later in this essay I shall base the statistical theory on a very strong generalization of the unique case rule. But it is at least possible that someone else should think the statistical theory prior, and use it to defend the long run rule.

The long run rule begins with a certain proposition. Call it the frequency proposition: all $S$'s are either $A$ or $B$, never both, and most often $A$. The long run rule is to be applied when experimental results show this to be true. The rule states that if the frequency proposition is known for some $S$'s, and one is guessing on a long sequence of $S$'s, the best policy is to guess $A$ every time.

So far as I know zero $A$'s in a long run of 200 or 2000 or 2,000,000

$S$'s is formally consistent with the frequency proposition. Anyway, the contrary has never been proved, so I shall not assume it. But zero $A$'s in a long run of $S$'s provide exceptional evidence against the frequency proposition. A coin falling tails 2000 times in succession is conclusive evidence that heads and tails do not have an even chance as outcomes from this coin; it conclusively shows that the coin is biased, or at least, was biased under past conditions of tossing. On any sound statistical theory, such evidence would justify rejecting the hypothesis of unbiasedness.

It is a standard statistical problem, to decide whether or not things like the frequency proposition should be rejected on the basis of given statistical data. Several theories, all pretty sound, draw the line between rejection and non-rejection. Let us suppose a man accepts one of these theories, and is about to guess on a long sequence of $S$'s. The arrangement of $A$'s and $B$'s in the sequence guessed on will be called the outcome of the sequence, since it so resembles the outcome of an experiment. Let our man evaluate his data in support of the frequency proposition. On any statistical theory, no matter how weak, and on any empirical data, no matter how strong, there will be possible outcomes which would justify rejecting the frequency proposition. No matter what the prior evidence and what the sound statistical theory, 2000 $B$'s in a long sequence of 2000 $S$'s would, together with the prior evidence, justify rejecting the frequency proposition.

Hence our man can divide possible outcomes of the sequence to be guessed on into two classes—those which would, given the prior data and a statistical theory, suffice to reject the frequency proposition, and those which would not. Let members of the latter class be called admissible. Aside from certain pathological kinds of prior data, it will happen that for any sound statistical theory, there will be more $A$'s in any admissible outcome than in any outcome which is not admissible.

The long run rule is to be applied on the basis of the frequency proposition. If, after the rule has been applied and guesses made, the frequency proposition must be rejected, the long run rule cannot be blamed. (If you follow the rule, bet on all horses owned by $X$, hear he owns $Y$, bet on $Y$ and lose but learn that after all $X$ does not own $Y$, you cannot blame your rule.) Hence the rule is culpable only if the outcome is admissible. It happens that amongst admissible outcomes, the maximum number of errors which could be incurred by the long run rule is less than the maximum number which might be incurred by any other viable rule. In this special sense the long run rule is guaranteed a lower upper-bound to error than any other rule. It minimizes the maximum possible error, and so may be called a minimax rule, though its rationale differs from what are more often styled minimax rules.

I think this whole argument very dubious. But perhaps it explicates what some people have in mind when they so gladly assent to the long run rule. I do not think it is fitting even to discuss the palpable lacunae in the above argument. It would be tedious to explore them when the long run rule is not even the subject of my present discussion. There seem to me three alternatives. This minimax argument may be patched up and accepted as the cardinal basis for the long run rule. Or there may be some argument which applied indifferently to the long run and unique case rules, and which, although justifying the long run rule, justifies the unique case rule independently, and so is not a long run justification of that rule. Or, it may happen that there is no further justification of the long run rule, and it must be accepted as a postulate. The second of these three possibilities will be discussed later. The third, if correct, virtually nullifies the long run justification of the unique case rule. I shall first explore the first possibility, and see how the minimax argument bears on the long run justification.

*The long run justification*

To return to the urn, I claim for the present that if certain frequency assumptions are in order, and one is guessing on a long sequence of draws with replacement, then, on the limited data, the best policy is to guess black every time. Does this entail a man should guess black if he is to guess only once, on a single ball, and never make another guess about this urn? Long run success in an hypothetical sequence never to be embarked on is little consolation if one should guess 'black' in this case, and white turns up.

Doubtless our business is not compassion but reason. Even if white turned up, he who guessed black made the most reasonable guess; 'black' was the hypothesis best supported. But does this follow from the long run rule? It might be urged that in a given state of information, there is but one reasonable guess. It is of no significance, the argument would continue, how many guesses are to be made. If reasonable and a million guesses to follow, then reasonable. In particular, reasonable if none to follow.

Tempting though it is, this argument is wrong. Guess of 'black' every time, in accord with the long run rule, is held reasonable because, in a certain sense, there is a guaranteed upper bound to error. This bound is less than that for any rival rule. But this reasonable making quality is not, as it were, transitive. If just one guess is to be made it is no longer true that a guess of 'black' has a smaller upper bound to possible error than any other rule. The upper bound is the same for any rule: exactly one error is always possible. No minimax argument will work here.

*The extended long run justification*

Even if you guess on but one draw from an urn, and even if our suitor guesses on only one letter, people may guess on many different things on similar frequency evidence. At a later and more sedate stage in his life, our suitor may guess on the sex of his unborn child knowing that, for genetic reasons, his family tends to girls. Hence it may be argued that he should adopt the long run rule as a life-long policy covering all the guesses he will make on things of which it is known only that things of that sort more commonly have one property than an alternative. He extends the long run justification, having in mind a long run of guesses on many things.

There is something resolutely unrealistic about the picture just described, but it seems to underlie the thinking of many people who accept the long run justification. It must be examined. Our suitor, it is supposed, accepts, perhaps regretfully, the possibility of going wrong on a letter or an urn, because he knows that, in the long run of all his guesses, the policy of guessing on the most frequent is the best policy. Hence the unique case rule is reasonable, for each time it is applied to work out the best guess, or the best supported hypothesis, our man is really just applying the long run rule in terms of all his guesses.

## Peirce's objection

Peirce describes a man who must guess on an issue of life and death, knowing only frequency data like that used in the unique case rule.† Long run success is not his immediate concern, but rather ensuring any run at all. No use to know that one rule is justified *if* he should survive; he wants first to survive. Or imagine our man choosing gruesomely between a bearable and an intolerable kind of death. Peirce grants that in such a case there is a reasonable guess, despite the absence of a future long run. He is a long run man, and faces the issue squarely. His hapless individual must embrace 'three sentiments' essential to logic and rationality. He must identify his aims with all men, think of himself as making one in the sequence of all man's guesses, and posit a prolonged existence for guessing sapient man. Thus his unique guess is justified not in terms of the long run of his experience, but in terms of the long run of human guessing. He makes the guess which, on behalf of all men, is prudent.

Doubtless there is some deep connexion between social aims and the rationale of the unique case rule, but I do not think it lies so plainly on the surface, or will ever be so high-minded. First of all, the best guess in the sorry situation above will be evident even to the selfish misanthrope who deliberately wishes evil on all mankind and hopes men's rationality will be confounded; it is evident to the wary philosopher who dissents from Peirce. Secondly, the most reasonable guess is evident even to a nuclear button pusher who, through our sovereign lack of reason, has been forced to guess, on mere frequency data, about an issue of life or extinction for all mankind. He cannot posit man's future. It would be foolish to demand that he merge his thought in hopes of cosmic rationality before he makes his guess.

The point of these fanciful tales is plain enough: it would be reasonable to guess black even if nobody made another guess. The fact that this would be the best guess under very different circumstances cannot be the justification for its being reasonable when no one will guess again.

† C. S. Peirce, *Collected Papers* (Harvard, 1938), II, p. 395.

## Other difficulties

Even though its author did not agree, Peirce's objection seems fatal to the extended long run justification. Other objections can of course be lodged. What sense, for instance, can be given to the earlier minimax argument when applied to a large number of different guesses, in different spheres of life, and whose consequences vary wildly? The form of the argument above assumes that the consequences of each guess are pretty well the same, but this is not true in life. Or again, there are various objections which call in question whether or not the conditions for applying the long run rule are really present, when the extended long run justification is stated more explicitly.† But Peirce's objection is so forceful that it should not be necessary to explore these other objections which can only ornament an already decisive reason for rejecting the extended long run justification.

## A possible axiom?

For all that has been said, it might be the case that every long-term guessing policy does, of necessity, coincide with the best guess in an unique case. For instance, the best guess about the ball in a unique case is 'black'; this is also the guess one would make in a long sequence of such guesses. Perhaps this should be accepted as an axiom. The axiom would state that all best long run and unique case policies do coincide, in some definite sense to be made precise. Without further considering grounds for the axiom, I shall prove that the 'axiom' is false. There is a counterexample.

Imagine that an urn is known to contain 100 balls of various colours. The frequency with which any colour is drawn (with replacement) from the urn is proportional to the number of balls of that colour in the urn. The exact constitution of the urn is unknown. But it is known that there are three possibilities: (*a*) 99 green balls, 1 red; (*b*) 2 green, 98 red; (*c*) 1 green, 99 red. One is to guess on the constitution of the urn after making one draw. There is only one urn. Green is drawn. What is the best guess? What the best supported hypothesis about the constitution of the urn?

† I. Hacking, 'Guessing by frequency', *Proceedings of the Aristotelian Society*, LXIV (1963–4), 62.

## A POSSIBLE AXIOM? 49

According to (a), there are many green balls in the urn, and a correspondingly large chance of drawing green; according to (b) and (c), the chance is slight. Everyone I have asked agrees that if he had to guess in this situation, he has no doubt that (a) is the best guess. The reasons given differ, but for the time being I take it as datum, that (a) is better supported than the other two possibilities.

Incidentally, this abstract case might have practical analogies. Perhaps a complex machine has been newly built for a laboratory. It is of course unique. Even a test on this machine is expensive. But one convenient test has just two possible outcomes, labelled $G$ and $R$. There are three hypotheses about the machine: (a) it is defective for every purpose; (b) it is defective for one purpose, not another; (c) it is not defective at all. On (a) the chance of $G$ is 0·99, on (b) it is 0·02, and on (c) it is 0·01. The test is run; no more can be afforded; the consequences of wrong decisions are equally grave whichever of (a)–(c) really holds. Which of (a)–(c) is best supported? If $G$ occurs, (a) is evidently best.

Those familiar with statistics need not make this point rest on untutored intuition. A standard statistical test of significance level 0·01 would reject (c) on the evidence of green; a test of size 0·02 would reject (b) on the same evidence. Hence any standard test of significance level 0·05, say, would certainly reject (b) and (c) leaving (a) not only as best supported hypothesis, but as the only tenable one. It also follows from Fisher's celebrated Principle of Maximum Likelihood that (a) is the best guess. But I do not think that principle can be taken for granted here.

Now suppose there are a great many such urns, or a great many such machines produced from different and independent manufactories. There is no reason to suppose the characteristics of one urn or machine are indicative of those of any other. Let there be the same three hypotheses about each urn: (a), (b) or (c). A ball is to be drawn from an urn; the constitution guessed on; then one passes to the next urn and repeats the business. Here is a long run of like guesses, each on a different urn. The sole aim is to be right as often as possible. What is the best policy?

One solution says we should use a 'mixed strategy', sometimes guessing (a), sometimes (b) when green is drawn. The actual ratio is determined by a minimax argument, though one different in kind from that described earlier. A computation shows that when green appears you should sometimes guess (*a*), sometimes (*b*), in the rather surprising ratio of 100:99. Which to guess on a particular occasion should be decided by tossing an almost fair coin. Likewise, if red appears you should guess sometimes (*c*), sometimes (*b*), in the same ratio. This strategem assures that no matter what the arrangement of constitutions of the urns, there is only a minute chance of guessing wrong more than about half the time, so long as you are guessing on a pretty long run. This minute chance diminishes as the run gets longer. It can be proved that no other policy is as good in this respect. Guessing (*a*) whenever green appears and (*b*) when red appears would, for example, be disastrous if all the urns were (*c*). But the minimax strategy would fare as well in that situation as in any other. Indeed other policies may have other pleasant characteristics, but one plausible school holds that a minimax argument is decisive in a case like this. The minimax strategem is the one which Braithwaite calls the most prudent.

Hence, in the long run, when green is drawn from an urn, it is reasonable, if not uniquely reasonable, to flip an almost fair coin, and guess (*b*) if it falls heads, (*a*) if it falls tails. But everyone, who has heard of this example, agrees that if he were guessing on a unique urn—perhaps on pain of life and death—it would be folly to flip a coin between (*a*) and (*b*) if green were drawn.

Hence there is at least one situation in which a guess, reasonable if a long run of like guesses is to be undertaken, is absurd if only one guess is to be made. It does not matter to this point whether the minimax long run policy is beyond a doubt the best policy. Personally, I prefer another, but admittedly one which is more of a gamble than the minimax policy. Someone who insists on minimizing the maximum possible risk is an over-cautious person who is unwilling to risk tiny loss for tremendous gain. But he is not silly. He would be silly if he tried to follow his policy on a unique guess when a green turned up.

The example of the three constitutions, of three kinds of machine, is not exactly like the single urn, nor does 'guess (*a*)' follow from the unique case rule. It is here used only to show that long run and unique case guesses need not always coincide. Any long run justification of the unique case rule, taking the coincidence as axiomatic, is certainly at fault.

*Other justifications*

One of the most ingenious and lucidly presented justifications of something like the unique case rule was propounded by Salmon; it is an adaptation of Reichenbach's famous 'vindication of induction'; one of the central defects in this mode of argument has been ably exposed by Salmon himself.†

Another, and older, possibility, takes a different tack. Could an equiprobability theory of probability, together perhaps with some form of principle of indifference, entail that if the frequency proposition is true ($S$'s more frequently $A$'s, etc.), and if $T$ is an $S$, then, on this data, it is more probable that $T$ is $A$ than that $T$ is $B$? I can only report that no such theory seems to have this consequence.

Indeed it is true that on equiprobability theories, 'more probable' is often defined as 'more worthy of confidence', or something of the sort. So one might state as a postulate that if the frequency proposition is true, it is more probable that $T$ is $A$ than that it is $B$. Perhaps many people have been feeling that my concern over the unique case rule is absurd: the unique case rule is right just because on the given data, it is most probable that $T$ is $A$. But unless there is a general theory about probability from which this follows, this is hardly a reason for the unique case rule. It is only a restatement of it. For if I understand this use of 'probable', it is supposed to convey that, on the given data, $T$'s being $A$ is better supported than $T$'s being $B$. Which is what the unique case rule says.

*Is the unique case rule true?*

The unique case rule might be rejected on two different grounds. Someone may discover a counterexample, and that will be that. It has not happened yet. But also, the rule might be rejected on philosophical grounds. Some who have been persuaded to long run justifications may insist that in cases where there is to be no long run of guesses, no guess based merely on frequencies can be called the best, and that no hypothesis is better supported than another. If long run arguments will not justify the unique case, nothing will, or so they may feel.

† W. C. Salmon, 'The Predictive Inference', *Philosophy of Science*, XXIV (1957), 180–90.

I cannot prove the unique case rule is not irrational: at any rate, if it be questioned, I cannot deduce it from other universally accepted truths. At most I can describe cases in which there will be no long run, and yet in which it would be right to use the unique case rule.

Suppose that an urn contains 99 black balls and 1 white one, and that in drawings with replacement, black is drawn correspondingly more frequently. Even if no one is to guess on anything again, and even if the urn and its contents are to be destroyed, 'black' is the best guess, and 'black' the best supported hypothesis about the next draw. An appeal to avarice may enforce this opinion. Suppose you were given a never-to-be-repeated offer. You must guess black or white, and lose a penny if you are wrong, gain £5 if you are correct. You prefer the fiver to the penny. Would you in these circumstances choose white? I think not. Can you seriously contend that your preference for black is entirely irrational?

Doubtless someone may insist that if there really is to be no future draw, and no more guessing by anyone, then no guess is better than any other, and no hypothesis better supported. He may say that it is habit that makes you suppose that if there were to be no long run, there would still be a reasonable guess. Probably he is right: the habit is a close relative of that mentioned by Hume, in connexion with induction. Whatever Hume may have thought, calling it 'habit' does not prove it irrational.

Other persons may say that, when black is drawn much more frequently than white, a guess of 'black' on the next draw is 'just what we do call' the best guess, and that the hypothesis of black is 'just what we do call best supported'. This is correct, though in an uninteresting way. What must be done is to state this facet of frequency in the long run as a postulate connecting support and chance. This has nothing much to do with the meaning of the English expression 'frequency in the long run'. The connexion would hold even if people had just noticed the physical property of frequency, and had no words to describe it, and were casting about for postulates to frame the property they noticed.

I believe that the unique case rule cannot be deduced from other universally accepted principles (except ones differing from it only in generality) and so it cannot be proved in that way. But the rule

## IS THE UNIQUE CASE RULE TRUE? 53

is correct. So it must indicate some fundamental fact about frequency in the long run. Someone trying to state postulates defining the property of frequency in the long run would have to include this fact. But what is the fact about frequency in the long run which is indicated by the unique case rule?

One could of course try to compose a precise statement of the unique case rule as it stands. But that rule is almost certainly not general enough. It is better to seek a principle which has wider application. In the next chapter this will lead us to the *law of likelihood*, which will serve as a foundation not only for guessing by frequency, but also for what is more commonly called statistical inference.

CHAPTER V

# THE LAW OF LIKELIHOOD

The unique case rule tries to state a connexion between frequency and support. In extending the idea it is useful to begin with a piece of reasoning based upon frequencies, but which the rule cannot validate. Imagine a textual critic editing a fragment from a Ciceronian manuscript. He knows the fragment is thirteenth century, but not whether it is faithful to its original. He must guess whether it comes from a reliable source or not. He notes a solecism of the sort seldom found in a classical author, but which all too often creeps in on the hand of a medieval copyist. On this slender data, he guesses his fragment is not to be trusted. Perhaps the inference can be schematized:

(1) every $X$ is either $Y$ or $Z$—every fragment that includes a solecism is either unreliable or reliable;

(2) the long run frequency of $X$'s among $Y$'s is greater than that among $Z$'s; therefore, lacking other information,

(3) the guess that this $X$ is $Y$ is better supported than the guess that this $X$ is $Z$.

In contrast, the unique case rule validates inferences of the following sort:

(1) every $X$ is either $Y$ or $Z$;

(2) the long run frequency of $Y$'s among $X$'s is greater than that of $Z$'s among $X$'s; therefore, lacking other information,

(3) the guess that this $X$ is $Y$ is better supported than the guess that this $X$ is $Z$.

Only the second premiss differs, but when schematized as above, it is plain that the inferences do differ. Yet the contrast need not be too great. We must not be entirely blinded by the beauties of logical form.

I shall argue that both inferences are validated by exactly the same fact about frequency and support. Perhaps it has never been denied that the unique case rule and the critic's guess have the same foundation, but I am sure most students assume the two are rather different. It is commonly believed that the rule is justified by long run success, but with the other sort of inference, most sober minds have balked at inventing a 'long run' of chance set-ups among which a man might be successful. So it has generally been supposed that the rule rests on different principles than the critic's guess. The preceding chapter proves the rule cannot be defended in terms of long run success. Hence the customary ground of distinction between the two modes of inference seems void. Indeed there are other dissimilarities. But the question is, whether the two can be validated by the same principle.

*Law of likelihood: vague statement*

In the second inference, connected with the unique case rule, why is 'This $X$ is $Y$' better supported than 'This $X$ is $Z$'? Without any paraphernalia of rules, one might simply say: because for 'This $X$ is $Z$' to be true, something must have happened on this particular occasion, which happens more rarely than what would have to have happened, for 'This $X$ is $Y$' to be true. This long-winded fact, it might be claimed, is just what provides better support for 'This $X$ is $Y$' than for 'This $X$ is $Z$'.

Putting the matter in this way is not free from objection, but it applies equally to the first and second mode of inference given earlier. I shall take this striking parallel as the clue to the fundamental fact which underlies both inferences. Nothing must hinge on the imprecise expression of the preceding paragraph, for it contains only a suggestion of a fact, not its accurate statement. The principle assumed seems to be more like this: *If one hypothesis h, together with data d, implies that something which is an instance of what happens rarely happens on a particular occasion, while another hypothesis, i, consistent with d, implies, when taken together with d, that something which is an instance of what happens less rarely happens on the same occasion, then, lacking other information, d supports i better than h.*

The italicized assertion could be used to validate both patterns of inference. In either case, let $d$ comprise the two premisses; let $h$ be 'This $X$ is $Z$' and $i$ be 'This $X$ is $Y$'.

It is now a task to make something precise out of the italicized assertion. Since the correct explication uses several new notions, it may be helpful to sketch them here and relate them to the matter in italics.

Instead of speaking vaguely of data $d$, we shall characterize a special sort of *statistical data*. This will be data about any set-up, stating a class of distributions, one member of which may be the true distribution of chances for trials of some kind. The data will also state a class of possible outcomes on some particular trial of that kind.

Instead of speaking vaguely of an hypothesis $h$ taken together with data $d$, we shall characterize a special sort of proposition. A *joint proposition* will be about trials of some kind on a chance set-up, and will state, jointly, (*a*) a single distribution of chances of outcomes for trials of that kind, and (*b*) that a specified outcome occurs on a designated trial of that kind. A joint proposition will be a kind of conjunction.

Instead of speaking vaguely of how $h$ and $d$ imply that something happens rarely, we shall characterize the *likelihood* of a joint proposition: a joint proposition states both an outcome and a distribution of chances; the likelihood of the joint proposition will be the chance of getting that outcome, if the distribution is the true one.

Finally, the *law of likelihood* will imply that if two joint propositions are consistent with the statistical data, the better supported is that with the greater likelihood.

The law of likelihood deals only in joint propositions, but when included in the logic of support, it has many consequences about other propositions as well.

## *Likelihood*

The concept of likelihood is due to Fisher. In his usage, it is essentially a term applicable to hypotheses about the distribution of chances from trials of a certain sort on a chance set-up. If the distribution is discrete, with a finite number of possible outcomes each with a chance of more than zero, the likelihood of an hypothesis, given that a trial has outcome $E$, is what the chance of $E$ would be, if the hypothesis were true. For instance, if a coin has been tossed twice, and fallen heads both times, then the likelihood that these tosses were independent with $P(H) = 0.7$, is just $0.49$; the likelihood that they were independent with $P(H) = 0.3$, is only $0.09$.

Likelihood does not obey Kolmogoroff's axioms. There might be continuously many possible hypotheses; say, that $P(H)$ lies anywhere on the continuum between 0 and 1. On the data of two consecutive heads, each of this continuum of hypotheses (except $P(H) = 0$) has likelihood greater than zero. Hence the sum of the likelihoods of mutually exclusive hypotheses is not 1, as Kolmogoroff's axioms demand; it is not finite at all.

In the first statement of the law of likelihood, it will be convenient to restrict attention to discrete distributions. This is done not only to avoid the less trifling mathematics which continuous distributions require, but also because, in any real experimental situation, there are only a finite number of possible outcomes of a measurement of any quantity, and hence only a finite number of distinguishable results from a chance set-up. Continuous distributions are idealizations. The rôle of the forthcoming law of likelihood may be better understood if this fact is kept in mind.

Recalling the textual critic, he must choose which of two hypotheses is best supported. According to one, something which happens rarely has happened; according to the other, something which happens less rarely has happened. Hence the latter hypothesis has, in the technical sense, greater likelihood than the former: if the latter is true, the chance of what happened is fairly large; if the former is true, the chance of what happened is slight.

Fisher generally uses likelihood as a predicate of hypotheses, in the light of data. Other writers have used it both as a predicate of hypotheses, in the light of data about outcomes, and of outcomes, in the light of data about hypotheses.† In at least some of his definitions, Fisher presents likelihood as if it were the predicate of an ordered pair, namely of a statistical hypothesis and an outcome; the likelihood is the chance of the outcome if the hypothesis is true.‡ This idea is followed here; a proposition stating jointly a statistical hypothesis and the occurrence of a particular outcome is called a joint proposition. Likelihoods will apply to joint propositions. The definitions are as follows.

A *simple joint proposition* is a proposition of the form, 'the distribution of chances on trials of kind $K$ on set-up $X$ is $D$; outcome $E$ occurs on trial $T$ of kind $K$'. It can be represented as a

---

† For instance, M. G. Kendall and A. Stuart, *The Advanced Theory of Statistics* (3 vol. ed.), II (London, 1961), p. 8.

‡ For instance, R. A. Fisher, 'On the mathematical foundations of theoretical statistics', *Philosophical Transactions of the Royal Society*, A, CCXXII (1922), 310.

sextet: $\langle X, K, D; T, K, E \rangle$—set-up, kind of trial, distribution; particular trial, kind of trial, outcome of the particular trial. Each side of the semi-colon represents a proposition; the simple joint proposition is a conjunction of these two.

For discrete distributions, the *likelihood* of the simple joint proposition $\langle X, K, D; T, K, E \rangle$, is what $P(E)$ would be, on trials of kind $K$, if the distribution were $D$. Or, briefly, $P(E)$ according to $D$. Until further notice discussion is limited to discrete distributions.

The statistical data which we shall employ can also be defined in terms of a similar notion. A composite joint proposition, or for short, a *joint proposition*, is one which states, 'the distribution of chances on trials of kind $K$ on set-up $X$ is a member of the class $\Delta$; outcome $E$ occurs on trial $T$ of kind $K'$'. Here $K'$ might be $K$, but need not be. A joint proposition is represented:

$$\langle X, K, \Delta; T, K', E \rangle.$$

According to our earlier custom of not distinguishing unit classes from their sole members, simple joint propositions are joint propositions.

Two joint propositions are *equivalent* if they entail each other. A joint proposition $\langle X, K, \Delta; T, K', E \rangle$ will be said to be *included* in another joint proposition $e$ just if there is some joint proposition $\langle X, K, \Delta'; T, K', E' \rangle$ equivalent to $e$, and such that $\Delta$ is included in $\Delta'$ and $E$ in $E'$.

These are the materials for our law of likelihood. They had better be contrasted with a similar idea in the literature. Barnard has devised what he calls the 'concrescence' of a statistical hypothesis and an experimental outcome.† Somehow this is not to be a mere conjunction of the two propositions; moreover the likelihood of a concrescence, when the outcome is known, is distinguishable from the numerically identical likelihood when the hypothesis is known. Our joint propositions, in contrast, are mere conjunctions, and can be treated by logic like any other conjunctions. Moreover, their likelihood is 'absolute', and not dependent on what is known. A likelihood as defined above is just a number attached to a certain kind of conjunction.

† G. A. Barnard, 'Statistical inference', *Journal of the Royal Statistical Society*, B, XI (1949), 115–39.

LIKELIHOOD 59

One peculiarity is worthy of mention. I shall sometimes speak of joint propositions serving as statistical data. A joint proposition asserts jointly a class of possible distributions, and the outcome of some particular trial. Normally one thinks of only the latter as data. When statisticians mention the former, they often call it part of the specification of the problem they are working on, and do not think of it as data at all. All the same, it serves as one of the datum points from which they argue, and so is part of their data. When we come to study the testing of statistical hypotheses we shall see that this kind of data is very important.

Finally, note that both conjuncts of a joint proposition might be practically vacuous. The proposition might assert merely that the true distribution is among the logically possible distributions, and that the outcome of a trial $T$ is the sure outcome.

*The law of likelihood for discrete distributions*

*If h and i are simple joint propositions and e is a joint proposition, and e includes both h and i, then e supports h better than i if the likelihood of h exceeds that of i.*

It will be noted that this law employs no non-logical terms which are not defined in terms of 'chance'. Hence it may be added to Koopman's underlying logic as a postulate to be used in the very definition of chance.

*Implications of the law*

It is worth observing a few easy consequences of the law within the logic of support.

(1) The unique case rule follows from the law, or, at least, its obvious explication does follow. Take a particular instance. We have a coin which may fall either $H$ or $T$, with $P(H) = 0.9$. In our abstract terminology, this means we have the statistical data:

On a simple trial from this set-up the distribution of chances of outcomes is $P(H) = 0.9$, $P(T) = 0.1$. On the next trial of this kind the outcome is either $H$ or $T$.

We compare the two simple joint propositions, $h$ and $i$.

$h$: The next simple trial from this set-up will have outcome $T$; the distribution of outcomes on simple trials is $P(H) = 0.9$, $P(T) = 0.1$.
$i$: The next simple trial will have outcome $H$; the distribution of outcomes is as stated in $h$ above.

Both $h$ and $i$ are included in the statistical data. The likelihood of $h$ is 0·1, and of $i$, 0·9. Hence by the law of likelihood, $i$ is better supported by the statistical data than is $h$.

This still does not tell us that $H$ is a better guess about the next trial than $T$. But it is a thesis of support that if $p\&q$ is better supported by $q$ than is $r\&q$, then $p$ is better supported by $q$ than is $r$. Now $i$ is the conjunction of part of the statistical data, with the proposition that the next outcome will be $H$. $h$ is similar. Hence by the thesis on support, the fact that $h$ is better supported than $i$ implies that *the hypothesis of $H$ on the next trial is better supported than the hypothesis of $T$.*

Hence the logic of support, together with the law of likelihood, entails just what the unique case rule entails. $H$ is better supported by the statistical data than is $T$. The above is a tedious proof of this obvious conclusion. Tedious or not, we have to show that such obvious truths do follow from our general law.

(2) The law also provides some inverse inferences. Take the example of the urn which may have one of three possible constitutions: (a) 99$G$, 1$R$; (b) 2$G$, 98$R$; (c) 1$G$, 99$R$. The chance of drawing $G$ or $R$ from the urn is in proportion to this ratio. $G$ is drawn. Which hypothesis about the urn is better supported on this data, (a), (b) or (c)?

The *simple joint propositions*, abbreviated, are

$h_1$: $G$ is drawn, and the distribution is as defined by (a);
$h_2$: $G$ is drawn, and the distribution is as defined by (b);
$h_3$: $G$ is drawn, and the distribution is as defined by (c).

The *statistical data* state that the true distribution is as defined by (a) or (b) or (c), and that green is drawn.

All three joint propositions are included in the statistical data. Their likelihoods are 0·99, 0·02 and 0·01 respectively. Hence by the law of likelihood, $h_1$ is best supported. Using the logic of support as in the preceding example, we infer that the hypothesis (a) is better supported, by the data, than the other two hypotheses.

Note that if $R$ has been drawn, (c) would have been best supported, but the contrast between the likelihoods of (b) and (c) would have been much less than that between those of (a) and (b), when $G$ is drawn. This suggests that the actual ratio of the likelihoods be a measure of the difference in support; if $R$ is drawn, (c) is better supported than (b) only in the ratio 99/98, while if $G$ is drawn, (a) is better supported than its nearest rival by the ratio 99/2. Tempting though this measure may be, it does involve difficulties.

(3) Finally a more interesting inference. Suppose a coin is tossed $n$ times, and that it falls heads on $k$ of these. It is known or assumed that the tosses are independent. Which hypothesis about $P(H)$ is best supported? In our reasoning, we use a fact demonstrated earlier: that the chance of getting $kH$ in $n$ independent trials is a function of $P(H)$, $k$, and $n$, according to the binomial theorem. Denote this chance by $B(k,n,P(H))$. Now consider an infinite set of joint simple propositions, all of the form,

$h_p$: On a particular sequence of $n$ tosses of this coin, the outcome contained exactly $k$ occurrences of $H$; the distribution of outcomes on trials consisting of $n$ tosses is exactly $B(k,n,p)$.

In this infinite set of hypotheses, $p$ ranges between 0 and 1. The statistical data state that $n$ tosses gave $k$ heads, and that the chance of this event lies between 0 and 1. Each hypothesis $h_p$ is included in this data. Hence the best supported is that with greatest likelihood. But it is easy to prove that $B(k,n,p)$ is a maximum when $p = k/n$. By the usual logic of support, we infer that, on the data of $k$ heads in $n$ tosses, the best supported hypothesis about $P(H)$ is that $P(H) = k/n$. If there are 1000 trials and 800 heads, the best supported hypothesis is $P(H) = 0.8$.

## Another version of the law

The law of likelihood can be expressed differently, and although it requires one more new concept before it can be stated, some readers may find this the more natural way of speaking. So far, likelihoods are numbers assigned to simple joint propositions in accordance with their conjuncts. They are 'absolute' likelihoods: the likelihood of a simple joint proposition is not altered by data. Now let us introduce a new notion, of the likelihood of a proposition given data. It will not differ much from the previous idea, but will make it possible to speak of the likelihood of many more propositions.

The definition proceeds by two stages. First, the likelihood of a simple joint proposition $h$ given the joint proposition $e$ which includes $h$ shall just be the 'absolute' likelihood of $h$. Secondly, an equivalence condition. If $d$ implies that the propositions $h$

and $i$ are equivalent, and if the likelihood of $h$ given $d$ exists, then the likelihood of $h$, given $d$, equals the likelihood of $i$, given $d$.

More briefly, our conditions are (1) if $d$ includes the simple joint proposition $h$, the likelihood of $h$ given $d$ is the 'absolute' likelihood of $h$; (2) if $h$ and $i$ are equivalent given $d$, then their likelihoods given $d$ are the same.

Now we can restate the law of likelihood: If the likelihoods of $h$ and of $i$ given $d$ do exist, *d supports h better than i if the likelihood of h given d exceeds that of i.* It is trivial to check that within the logic of support this law is logically equivalent to the preceding form of the law of likelihood. The reader may care to run through the three examples given earlier, using this new form of the law rather than the old one.

*History: maximum likelihood*

Some questions about the law of likelihood are best discussed in an historical perspective. The law has many antecedents. Since statisticians have never expressed themselves in terms of support, and since logicians of support have seldom studied statistical methods, nothing quite like the present law has ever been stated before. It must be distinguished from the most famous method of *estimation*. Fisher's *Principle of Maximum Likelihood* implies that the best estimate of a character of a chance set-up is that with greatest likelihood. Hence, knowing only that $H$ has occurred $k$ times in $n$ independent trials, the best estimate of $P(H)$ is $k/n$. But this is not a direct application of the law of likelihood; conversely, the claim that $k/n$ is the best estimate does not entail $k/n$ is the best supported hypothesis about $P(H)$. For as was remarked two chapters ago, whether or not an estimate is good or bad may depend on the purpose to which it will be put; it certainly depends in some way on the nearness of the estimate to the true value. These factors seem irrelevant to questions of support. Sometimes the best estimate and best supported hypothesis do coincide, but, as has already been proved by a counterexample, this coincidence is by no means necessary. More interesting examples will be treated in the chapter on point estimation.

From a philosophical point of view there is a more fundamental distinction between the law and Fisher's principle. Any principle of estimation must be judged according to its qualities, in particular, its ability to be near that which it is used to estimate. It must be proved that a principle has these qualities, or discovered where it does possess them, and when they are lacking. It is true that even today a few writers take Fisher's principle as axiomatic, but Fisher never did. Others had used the principle before; his contribution was to state it outright and discover the qualities which make it desirable. His success or failure is not to be questioned here; all that matters is that his principle is something to be proved or disproved. The law of likelihood, however, does not seem open to proof in the same way. Support is a more fundamental notion than being 'near' to a true value. The principle of maximum likelihood might be deduced from criteria for what makes a good estimator, but no similar criteria about support entail the law of likelihood. The law expresses a fundamental fact about frequency and support whereas Fisher's principle, if true at all, must be deduced from other facts.

The law of likelihood has received little attention from statisticians because it is hardly ever of practical importance. I suppose the practical man wants an estimate of the true value of something in accord with his needs, and not an abstract statement that something is better supported than something else. In early writings where likelihood is important, it is often impossible to tell whether the author has estimation or support in mind. A lucid discussion by Daniel Bernouilli provides a fine example of this difficulty, and is especially useful since its recent republication makes it generally available.†

Like so many early writers, Bernouilli examined the problem of reconciling discrepant measurements of the sort inevitable in a physical science like astronomy. Before his time most people thought obviously wild observations should be discarded, and the rest averaged. Bernouilli thinks taking measurements is like making trials on a chance set-up. The distribution of chances of various measurements depends on the true value of what is being measured. The further away is a measurement from the true value, the smaller the chance that someone will make it. Given a set of

---

† 'The most probable choice between several discrepant observations and the formation therefrom of the most likely induction', translation by C. G. Allen of the essay from *Acta Acad. Petrop.* (1777), appearing in *Biometrika*, XLVIII (1961), 3–13.

measurements, the problem is to guess at the true value. Bernouilli makes an assumption about the shape of the distribution curve. He then takes as most probable that hypothesis about the true value which has the greatest likelihood: 'I think that of all the innumerable ways of dealing with errors of observation one should choose the one which has the highest degree of probability for the complex of observations as a whole'.

It can be shown that best estimates and best supported hypotheses do coincide for the case which Bernouilli has in mind. He, writing in 1777, could not be expected to distinguish the two concepts. His arguments may be construed as applying either to the law of likelihood, or to the principle of maximum likelihood. I think they bear more on the former; Kendall, prefacing the translation of Bernouilli's article, has no doubt that they pertain to the latter. It does not matter much.

Bernouilli keenly observes of his argument, 'all this I would wish to have weighed in the balance of metaphysics rather than mathematics'. He does not try to deduce his theory from known principles, but tries to make it plausible as incorporating a fundamental postulate. Though he does this by trying to make it self-evident, his arguments are of interest.

Notably, he differs from nearly all his successors in having no qualms about comparing direct and inverse arguments. He cites examples of direct inferences about dice, and tries to show that even they rely on comparing likelihoods. He compares his problem to that of finding the bull's eye at which a fine archer was aiming, when we know only the points at which his arrows hit; he argues that the bull must lie where the hits are most dense. This is equivalent to taking the hypothesis of greatest likelihood. Never does he deduce his principle, but continually argues by analogy, that willing slave of metaphysics. Today we are better equipped. If concerned with the principle of maximum likelihood, Fisher's discoveries help in deducing desiderata; if developing the law of likelihood, one may show as in future chapters how it leads to many conclusions which are probably true, and which apparently could not be true if the law were false.

*History: the likelihood principle*

Recently several writers have canvassed a 'likelihood principle', which states, roughly, that in assessing statistical hypotheses in the light of experimental evidence, only likelihoods count. This idea may underly much of Fisher's work. An early statement of the principle is the most clear: 'What, after all, *is* a simple statistical hypothesis? What does it do for us? It enables us to attach a number to experimental results—the likelihood of such results, on the hypothesis in question. The connexion between a simple statistical hypothesis $H$ and observed results $R$ is entirely given by the likelihood or probability function $L(R|H)$'—the likelihood conceived as a function of results $R$ and hypothesis $H$. 'If we make a comparison between two hypotheses, $H$ and $H'$, on the basis of observed results $R$, this can be done only by comparing the chances of, getting $R$, if $H$ were true, and those of getting $R$, if $H'$ were true.'†

The principle says that only likelihoods count: 'Given the likelihood function in which an experiment resulted, *everything* else about the experiment...is irrelevant.'‡ Another author is a little more specific: everything else is irrelevant to the 'evidential meaning' of the outcome of an experiment.§

The likelihood principle does not entail the law of likelihood since it does not say how likelihoods are relevant. The two are consistent; the one may suggest the other. Since the law makes bolder and more specific claims it is to be preferred to the principle. But the scope of the principle may be broader; it may be construed as a rule of thumb for designing tests of statistical hypotheses, whereas the law does not say, explicitly, anything about testing hypotheses. The principle will not assume any importance until our discussion of tests. Even there, it seems that specific criticism of particular theories of testing is better than blanket exclusion on principle of some theories which would use data other than likelihoods. But

† G. A. Barnard, review of *Sequential Analysis, Journal of the American Statistical Association*, XLII (1947), 659.
‡ L. J. Savage, 'The foundations of statistics reconsidered', *Proceedings of the Fourth Berkeley Symposium on Mathematical Statistics and Probability*, 1 (1961), 583.
§ A. Birnbaum, 'On the foundations of statistical inference', *Journal of the American Statistical Association*, LVII (1962), 269–306.

such matters are of necessity postponed until our two chapters on testing.

Modern champions of the likelihood principle claim to find its origin in Fisher. His reasons for accepting it may have resembled our discussion climaxing in the law of likelihood. But contemporary exponents, of whom Savage has been the most notable, find the principle compelling for reasons which derive from a sophisticated form of the studies pioneered by the eighteenth-century polymath, Thomas Bayes. These reasons will be exploited in due course, but for the present, none can furnish direct support for the law of likelihood. We shall treat the likelihood principle after expounding Bayes' theory. For the present, it is merely pleasant that a principle, in some ways related to our law, should be enunciated by writers with quite different preconceptions, and who do not profess much explicit interest in the physical property we are trying to analyse.

*Continuous distributions*

The likelihood of joint propositions has been defined for discrete distributions. When defined, one joint proposition is asserted to be better supported than another if it has greater likelihood. This condition is equivalent to another: the ratio between the likelihoods should be greater than one. The likelihood ratio will be defined for continuous distributions, and serve for a general statement of the strong law of likelihood.

There are rigorous definitions of likelihood ratios. They are best expressed in measure theory, the mathematical theory of the measure of sets of points. We do not need a new or better definition, but a description of why the measure theoretic definitions are satisfactory tools of inference. This can be done with little rigour, so that the course of ideas will be evident to anyone. Those familiar with measure theory can fill in the details. But I do not think this is entirely an exercise in popularization. On the contrary, only when we attend to the original meaning of our theoretical constructs shall we make sound inferences.

Waiting times at a telephone exchange were used, a few chapters ago, to introduce continuous distributions. The waiting times between two calls consecutively arriving at an exchange is the time between one call and the next. Here is a chance set-up. The possible

CONTINUOUS DISTRIBUTIONS 67

results of a trial are the possible waiting times. If time is conceived as infinitely divisible, there is a continuum of possible waiting times. It will be recalled how the distribution of chances of these outcomes may be represented by a cumulative distribution function, like that whose curve was drawn on p. 18 above. If the cumulative function is $F$, then $F(t)$ is the chance of a waiting time being less than or equal to $t$. It will also be recalled that suitably differentiable continuous distribution functions are expressible as integrals of what have been called *density functions*. In many interesting cases, where $F$ is the cumulative function, there will be a density function $f$, such that $F(t)$ is the area under the curve $f$ up to the point $t$, that is,

$$F(t) = \int_0^t f(x)\,dx.$$

Let $h$ and $i$ be two rival hypotheses, $h$ stating that the cumulative function is $F$, and $i$, that it is $G$. Given the result of a sequence of independent trials, how is one to determine whether $h$ or $i$ is better supported?

To begin with, consider a single observed waiting time, $A$, and let this be the sole data adjudicating between $h$ and $i$. $F(A)$ and $G(A)$ represent the chances of getting a waiting time less than or equal to $A$, according as the true distribution is $F$ or $G$.

$$F(A) - F(A - \epsilon)$$

is the chance of getting a waiting time between $A$ and $A - \epsilon$ when the true distribution is $F$. But if $F$ is continuous, this converges on zero as $\epsilon$ decreases. Hence if likelihood were defined in analogy with discrete distributions, the likelihood of outcome $A$ would be zero. There is zero chance of getting exactly $A$.

But no actual measurement can entail that the waiting time is exactly $A$. Ordinary measurements might be to the nearest second, extraordinary ones to the nearest millisecond. A measurement entails at most that the observed waiting time lies within a narrow interval. The true experimental result is not a point $A$, but an interval around $A$. Intervals do typically have positive likelihood: there is more than zero chance of getting an outcome in the interval between $A - \epsilon$ and $A + \epsilon$. When the true distribution is $F$, this chance is $F(A+\epsilon) - F(A-\epsilon)$. When it is $G$, the chance is

$G(A+\epsilon)-G(A-\epsilon)$. So if the outcome lay in this interval around $A$, the likelihood ratio between our two hypotheses would be

$$\frac{F(A+\epsilon)-F(A-\epsilon)}{G(A+\epsilon)-G(A-\epsilon)}. \tag{1}$$

Unfortunately it is not usually possible to state the exact interval around $A$ in which the true value must lie. This is itself a problem in chances. Hence it will not do to work out the likelihood ratio with any given $\epsilon$ in mind. We need a definition of likelihood ratio which will be approximately correct for any small value $\epsilon$. Under very general conditions this is perfectly possible. When $\epsilon$ is small and $F$ and $G$ are in a certain sense well behaved, the likelihood ratio (1) will be about the same for different values of $\epsilon$, and will converge to a limit as $\epsilon$ decreases. The likelihood ratio will be defined only when this limit exists. And here is where density functions come in: the likelihood ratio (1) will, as the next section explains in detail, converge on the ratio of the value of the density functions at $A$. This ratio is important in inference not because it represents the ratio between the chance of getting exactly $A$, when $F$ is correct, and the chance of getting exactly $A$, when $G$ is correct. It does no such thing. It represents, roughly, the ratio between the chance of getting an outcome about $A$, when $F$ is correct, and an outcome about $A$, when $G$ is correct.

*Experimental densities*

So much for the rationale behind likelihood ratios for the continuous case. A definition of likelihood ratios for well-behaved functions is best explained in terms of densities. This is familiar to statisticians but is worth going through from a very naïve point of view.

In Fig. 3 let $F$ and $G$ be two cumulative distributions, and $f$ and $g$ be their density functions. This means that the value of $F(t)$—the length of the line $a$ on the right—is just the area under the curve $f$ up to the point $t$. The chance of getting a result between $t$ and $t+\epsilon$ is $F(t+\epsilon)-F(t)$. This is the difference between lengths $a$ and $b$ on the right, and hence equal to the shaded area of the left-hand diagram. Moreover, the likelihood ratio

$$\frac{F(t+\epsilon)-F(t)}{G(t+\epsilon)-G(t)}$$

# EXPERIMENTAL DENSITIES

is the ratio between two such areas. As $\epsilon$ becomes small this evidently approximates to the ratio of $f(t)$ to $g(t)$. So the likelihood ratio around $t$ is approximated by the ratio of the densities at $t$.

I have spoken as if density functions were unique, but this is a fiction. $f$ is the derivative of $F$ with respect to $t$. The derivative with respect to $t^2$ could differ, and still seem to be a 'density' function. But if you try to repeat the argument of the preceding section using $t^2$ instead of $t$, you get a nonsense. So we shall insist that the density function for defining likelihood ratios should be the *experimental density*. This is to be defined as the derivative of the cumulative function with respect to the quantities actually

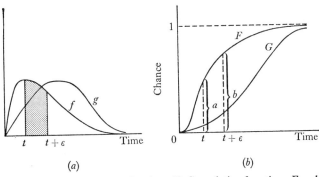

Fig. 3. (a) Density functions $f$ and $g$. (b) Cumulative functions $F$ and $G$.

being measured. Henceforth we assume that a specification of a kind of trial includes a specification of what quantities are measured in order to determine the experimental result. Then if

$$\langle X, K, D; T, K, E \rangle$$

is a joint proposition, $D$ a continuous and differentiable distribution, and $E$ a result, the experimental density of $\langle X, K, D; T, K, E \rangle$ is the value of the experimental density function of $D$, at the point $E$.

For a slightly less informal definition of experimental densities, let the result of a trial of kind $K$ be represented by $z_1, \ldots, z_n$, an ordered set of real numbers. Thus we consider a cumulative distribution function $F(z_1, \ldots, z_n)$. But the $z$'s might not be measured directly. Let $x_1, \ldots, x_m$ be the quantities which are actually measured at a trial of kind $K$, and which vary from trial to trial.

Then the *experimental density function* for trials of kind $K$ shall be the partial derivative,
$$\frac{\partial^m F(z_1, \ldots, z_n)}{\partial x_1, \ldots, \partial x_m}$$
expressed as a function of $z_1, z_2, \ldots, z_n, x_1, \ldots, x_m$. A completely formal definition is given only with the aid of measure theory.†

*Likelihood ratios*

Likelihood ratios will be defined only for two cases, albeit very rich ones. Once again, a more general measure-theoretic definition can be given, which will include in a single rubric both the cases given here. But the two cases I discuss are the only ones for which there is, at present, an entirely clear experimental interpretation.

When the likelihoods of both
$$\langle X, K, D: T, K, E \rangle \quad \text{and} \quad \langle X, K, D'; T, K, E' \rangle$$
are greater than zero, their likelihood ratio is defined as the ratio of their likelihoods. And in the continuous case, when both $E$ and $E'$ are results, the likelihood ratio of the two propositions is the ratio of their experimental densities. This elementary notion suffices for a general statement of the law of likelihood.

*The law of likelihood*

The law will be defined only for those cases in which likelihood ratios exist. This does cover all discrete distributions and every other distribution extensively studied or for which any purpose has been found. The law of likelihood: *If h and i are simple joint propositions included in the joint proposition e, then e supports h better than i if the likelihood ratio of h to i exceeds 1.*

This I venture as the explication of the thesis, 'if $p$ implies that something happens which happens rarely, while $q$ implies that something happens which happens less rarely, then, lacking other information, $q$ is better supported than $p$'. Perhaps some degree of obviousness could be claimed for the law of likelihood, but this is not terribly important. What matters is whether it demonstrably provides a basis for much statistics known to be true, without validating what is false.

† Along the lines of P. R. Halmos and L. J. Savage, 'Application of the Radon–Nikodym theorem to the theory of sufficient statistics', *Annals of Mathematical Statistics*, xx (1949), 225–41.

## Another version

For chance set-ups with only finitely many possible results we were able to state the law of likelihood in two distinct versions. One used the idea of the 'absolute' likelihood of a simple joint proposition, while the other considered likelihoods of propositions given data. Naturally we can do the same thing for the continuous case.

The *likelihood ratio of h to i given data d* is, as before, defined in two stages. First, if the joint proposition $d$ includes the simple joint propositions $h$ and $i$, then the likelihood ratio of $h$ to $i$, given $d$, is simply the 'absolute' likelihood ratio of $h$ to $i$. Secondly, if $d$ implies that $h$ and $i$ are equivalent, and if the likelihood ratio of $h$ to $j$ given $d$ does exist, then so does the likelihood ratio of $i$ to $j$ given $d$, and it equals the likelihood ratio of $h$ to $j$, given $d$.

That is to say: for simple joint propositions included in the joint proposition $d$, likelihood ratios given $d$ are absolute likelihood ratios. And for any propositions whatsoever if $h$ and $i$ are equivalent given $d$, their likelihood ratios to any $j$, given $d$, are just the same, if they exist at all.

The law of likelihood may be restated: *d supports h better than i whenever the likelihood ratio of h to i given d exceeds* 1.

## The Normal distribution

It is no part of this essay to attempt a study of particular statistical distributions. But the distribution called *Normal* in English and *Gaussian* in other languages is so ubiquitous in statistics that it cannot fail to be useful for illustrating particular points. So it will be convenient to set down for the record the properties of Normal distributions; they will not be used in the central argument of anything that follows, but will occasionally be referred to when producing examples.

Normal distributions are among those whose density function is the familiar 'bell-shaped curve'; thus the density and cumulative functions (Fig. 4) give an idea of the shape of the distribution. Such distributions were first investigated by de Moivre at the beginning of the eighteenth century. He proved that if we consider binomial distributions based on more and more trials, the distribution of chances more and more closely approximates to a Normal curve. Gauss studied Normal curves in part because of his interest in the theory of errors; if one conceives of an error on a measurement as produced by a large number of independent events binomially distributed, one is led to suppose the distribution must be more or less Normal, in virtue of de Moivre's theorem. In the nineteenth century the great demographer Quetelet became obsessed by the fact that a host of empirical phenomena seem correctly described by bell-shaped curves. In the Gaussian tradition, the mathematician Lexis elaborated the mathematical properties. The geneticist Galton and the statistician Karl Pearson developed an almost baroque theory of approximation to Normalcy.

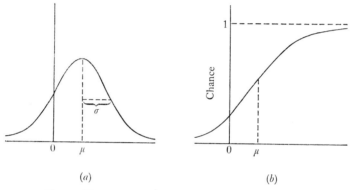

Fig. 4. The Normal distribution. (*a*) Density function.
(*b*) Cumulative function.

When Gossett and others initiated a more precise study of sampling from populations, it appeared that most familiar methods of sampling lead to curves which are at least roughly Normal, and which become increasingly Normal as the size of the sample increases. Meanwhile abstract probability theorists proved many theorems like that first one of de Moivre's. Recently these have been absorbed into yet more general limit theorems where Normalcy is no longer a cardinal feature but at most an incidental by-product. The whole history of mathematical statistics could be written under the head of the Normal family of distributions.

The actual form of a Normal curve can be given in terms of just two parameters. One is the *mean*, or average value of the distribution, marked by $\mu$ on Fig. 4. The other is a measure of the dis-

## THE NORMAL DISTRIBUTION

persion, or width of the curve about the mean. Many measures would do, but that usually taken is called the variance; the variance $\sigma^2$ may be thought of as the average value of the square of the distance of results from the mean. The analytic form of the curve is not something we shall much need in the sequel, but for the record, the density function has the form

$$\frac{1}{\sqrt{(2\pi)}\sigma} \exp\left[-\frac{1}{2}\left(\frac{x-\mu}{\sigma}\right)^2\right].$$

Now suppose you are given data $d$: on independent trials of kind $K$ the distribution of chances is Normal, of unknown mean and variance; $n$ trials of kind $K$ have been made, with results $x_1, x_2, ..., x_n$. On this data, what is the best supported hypothesis about the mean of the distribution? It is gratifying that the law of likelihood gives the result which has traditionally been accepted; the best supported hypothesis is, that the mean for this distribution of chances is just $(1/n)(x_1 + x_2 + ... + x_n)$. We cannot draw similar comfort from what, according to the law of likelihood, is the best supported hypothesis about the variance. For there have traditionally been a number of different answers and there has been no agreement on how to settle which is correct. So if we conclude by accepting the law of likelihood, we shall accept its recommendation for the variance of the distribution, but this recommendation cannot be used as corroboration for the law of likelihood. Nobody yet knows for certain which recommendation is best.

CHAPTER VI

# STATISTICAL TESTS

By rejecting a statistical hypothesis I shall mean concluding that it is false. On what statistical data should this be done? Braithwaite thought the matter so crucial that he tried to state the very meaning of 'probability statements' in terms of rules for their rejection. We shall examine his ideas later. First we must establish when evidence does justify rejection. To do so, it need not entail that the hypothesis is false. But what relations must it bear to the hypothesis?

Perhaps rejection covers two distinct topics. There have been many debates on this point, and it cannot be settled before further analysis. But a warning may be useful. An hypothesis may be rejected because of the evidence against it. This is my main subject. But situations can arise in which it is wise to reject an hypothesis even though there is little evidence against it. Suppose a great many hypotheses are under test. A good strategy for testing is one which rejects as many false and as few true hypotheses as possible. The best strategy might occasionally entail rejecting hypotheses even though there is little evidence against them. This sounds implausible, but examples will be given.

There is no general agreement on whether rejection should be studied in terms of evidence or strategies. I do not want to prejudge the issue. But I shall begin with examples in which an hypothesis should be rejected because of the evidence against it. I shall not begin with examples in which a great many similar hypotheses are under test. The logic of the two may be the same, for all that has been proved. But I shall not begin by assuming it.

The forthcoming discussion is, as usual, very academic. It concerns the relation of statistical hypotheses to statistical data. Generally one has all sorts of data bearing on an interesting statistical hypothesis, far more than merely statistical data. Hence one's problem is generally more complex than any to be discussed in this chapter. Here I deal only with data which may be precisely evaluated, and whose evaluation is peculiar to statistics.

The demand for precision must not be misconstrued. It would be wrong to expect an absolutely sharp line between those hypotheses which should be rejected on some data, and those which should not. One can draw such a line for no hypothesis whatsoever, statistical or otherwise. One may only compare the stringency with which different hypotheses have been rejected. There might be a scale, with 'very stringent' at one end and 'not very' at the other, but even this is more a matter for general considerations than for mathematics.

It is natural to think of the theory of rejection as part of decision theory, but for reasons given earlier I wish to avoid decision theory as much as possible. It should be possible to compare how different pieces of evidence justify rejecting any hypotheses without mentioning the expenses of correct or incorrect rejection. Indeed both the utility connected with an hypothesis, and the prior, non-statistical, information about it will affect the stringency with which one would reject an hypothesis. But it should be possible to study stringency before studying how stringent one should be in particular cases.

*Arbuthnot's test*

The first published test of a statistical hypothesis will illustrate many points. I refer to 'an Argument for Divine Providence taken from the constant Regularity of the Births of Both Sexes' which John Arbuthnot communicated to the Royal Society in 1710.† His ideas were in the air at the time. Indeed statistical hypotheses had been rejected long before: according to a traditional view, the study of probability began when de Méré inferred from tossing dice that '9' and '10' do not have equal chances from a pair of fair dice, and asked Pascal to explain the paradox. But Arbuthnot seems to be the first man to publish the reasoning behind a statistical inference.

His paper studies the hypothesis that it is an even chance, whether a child be born male or female. He takes as datum the register of births for the city of London. These go back 82 consecutive years. On every recorded year more boys were born than girls.

† John Arbuthnot, *Philosophical Transactions of the Royal Society*, XXIII (1710), 186–90.

Arbuthnot argues that if there is an even chance for male and female births, the distribution of births should be like outcomes from a fair coin. I shall use the term 'male year' for a year in which more boys are born than girls. Then, on the hypothesis of even chance, Arbuthnot thinks the distribution of male and female years should also be like the distribution of outcomes from a fair coin. He does not prove this but assumes it. He then tests his original hypothesis by testing a new hypothesis $h$: The distribution of chances of male and female years is binomial with $P(M) = 1/2$. In what follows I am concerned solely with his test of $h$. Passage from $h$ to the original proposition about male births is rather tricky, and beyond Arbuthnot's powers.

If $h$ were true, there would be only a minute chance of getting 82 male years in a row: $(1/2)^{82}$, 'which will be found easily by the table of logarithms to be 1/4836000000000000000000000'. Moreover if the surplus of male births per year prevailed 'in all ages, and not only in London, but all over the world (which 'tis highly probable is Fact...)' the chance of this perennial surplus would, on $h$, 'be near an infinitely small quantity, at least less than any assignable fraction'.

Arbuthnot seems to reason: an event has happened in London, as reported in the registers. If $h$ is true, the chance of that event happening is minute, nay, events of that sort would virtually never happen. So he rejects the hypothesis. Let us suppose the argument rests here, and does not rely on what Arbuthnot thinks probable, that more males are born than females in every year, in every clime.

In addition to rejecting $h$, and rejecting the hypothesis of equal chance of male and female births, Arbuthnot also concludes that 'it follows that it is art, not Chance, that governs' in the distribution of sexes. Hence the title of his paper. It is conceivable that this dilettante, amateur probabilist, physician to Queen Anne, buffoon, friend and fellow satirist of Swift, had tongue in cheek. But reading him straight he seems to have thought that if chance did govern, and if a statistical as opposed to a purely causal account were appropriate, it would be so only if the chances of male and female births were equal. Perhaps this stems from the idea that chances depend on equally possible alternatives. The only statable and relevant partition of the alternatives, as far as Arbuthnot could see, would be $M$ and $F$, giving equal chances to each; since this must be rejected, so must the hypothesis of chance. I can only guess. It is notable that Arbuthnot's argument for divine providence was instantly adopted by theologians and preached, midst more familiar nonsense, from the pulpits of Oxford and Munich for the rest of the century.†

Arbuthnot's argument also roused controversy amongst serious men, including the great probabilists de Moivre, s'Gravesande, Montmort and N. Bernouilli. The first three supported him but the last thought Arbuthnot quite wrong. Perhaps there was no very real dispute. Bernouilli was concerned to refute the conclusion of 'Art, not Chance' and did so conclusively; the others defended earlier steps in the argument. Bernouilli, for instance, says that a binomial distribution with odds of 18:17 on males perfectly accords with the data. Others say this misses Arbuthnot's point. de Moivre sanely concludes that the hypothesis of even chances is excluded, that odds of 18:17 are an acceptable analysis of the data, and that the odds being exactly this is no less than providential.‡

How sound is Arbuthnot's reasoning? I refer only to his rejection of $h$ on the basis of the registers, and neither to his inference to the chance of a child being born male, nor to his joke about providence. He is surely right to reject $h$ on the data, but why? He proceeds, I repeat, by proving that what has happened would happen extremely rarely, if $h$ were true. For simplicity, let us suppose that the registers show for each of the 82 years, only whether it is male or female. Arbuthnot rejects $h$ because the likelihood of $h$, together with the data, is too low, namely $(1/2)^{82}$.

Unfortunately it seems that with Arbuthnot's chain of reasoning we ought to reject the hypothesis no matter what happens. Imagine that the registers had reported the first year to be female, the next male, the next and the next female, and so on, in some definite order, giving 41 male years and 41 female ones. Here, we

† W. Derham, *Physico-Theology; or a demonstration of the being and attributes of God from the works of Creation* (London, 1713). J. P. Süssmilch, *Die Gottliche Orndung* (Berlin, 1741).

‡ P. R. de Montmort, *Essai d'Analyse sur les Jeux de Hazards* (2nd ed., Paris, 1714), pp. 373 f. and 388–93. A. de Moivre, *Doctrine of Chances* (3rd ed., London, 1756), pp. 251–4. W. J. s'Gravesande, *Oeuvres Philosophiques et Mathématiques* (Amsterdam, 1774), pp. 221–48. Letters from N. Bernouilli, dated 1712 and 1713, are printed in both s'Gravesande and Montmort. De Moivre's opinions were first published in 1713.

may say, is a result in accord with $h$: but the chance of this result is not different from that of the actual data, namely $(1/2)^{82}$. And so for every possible result. The likelihood of $h$, together with any possible result, is always $(1/2)^{82}$. If, as Arbuthnot seems to assume, we should reject whenever the likelihood is low enough, we should reject every possible result.

Common-sense finds this conclusion foolishly sophistical. According to $h$, the chance of getting exactly 41 male years—in some order or other—amongst 82 is excessively greater than the chance of 82 years, all male. Indeed, but Arbuthnot's data would not have been '41 male years and 41 female, in some order or other'. It would have been 82 years, arranged in some definite order. What entitles us to contract our data, and consider proportion and not order? It is obviously right to make just this contraction, but why?

Such questions demand a more explicit statement of Arbuthnot's test. His data consist of the registers which, we suppose, tell for each year whether or not more males were born than females. The possible results might range all the way from '82 female years' to '82 male years' with every permutation in between. We must consider a subclass of these results, and reject $h$ if and only if the actual result lies in this class. Let us call this class $R$, short for rejection class. If the observed outcome is contained in $R$, reject; do not otherwise. What is $R$ to be?

Probably Arbuthnot's intuition, which many will share, is that $R$ should be chosen so that if $h$ is true, the chance of a result being in $R$ is very small. Crudely, reject $h$ if what is observed would happen rarely if $h$ were true. This choice of $R$ has the advantage that if $h$ is true, there is very little chance of rejecting it.

The chance of $R$, on the assumption that $h$ is true, is nothing but the likelihood of a joint proposition, that from a certain chance set-up (the sequence of years) there is a certain result (being in $R$) on a certain kind of trial (observations lasting 82 years) when the distribution of chances of outcomes is as described in $h$. For brevity, represent this joint proposition by the ordered pair, $\langle R, h \rangle$. Arbuthnot would choose $R$ so that the likelihood of $\langle R, h \rangle$ is very small.

Unfortunately, in Arbuthnot's problem, with the stated data, there are a multitude of ways in which one might so choose $R$.

For any possible result against which one may conceive a prejudice, one may choose a class $R$ so that the result in question is a member of $R$ and the likelihood of $\langle R, h \rangle$ is as small as you please (down to $1/2^{82}$ anyway). So for all this machinery of rejection classes we are not yet better off than before.

Despite the numerous tests developed by Laplace both in isolated papers and his great *Théorie Analytique*, even he says little about how to choose the kind of results on which to base a test, or why to abstract from data about the order and proportion in male years, and treat only of proportions. In many papers he is too concerned with his supreme mathematical analysis to worry about application; in problems connected solely with applications, he generally argues like Arbuthnot, albeit with more powerful tools. But as so often happens in his writings, an informal aside solves our problem. To quote a famous passage,

> On a table we see letters arranged in this order, CONSTANTINOPLE, and we judge that this arrangement is not the result of chance, not because it is less possible than the others, for if this word were not employed in any language we should not suspect it came from any particular cause, but this word being in use amongst us, it is incomparably more probable that some person has thus arranged the aforesaid letters than that this arrangement is due to chance.†

The moral is plain. Rejection of the hypothesis of chance is not based solely on the fact that, if it were true, the name 'Constantinople' would very seldom be spelled out, but also involves the existence of another hypothesis according to which the spelling out would occur much more often.

Let us apply this insight to a trivial form of Arbuthnot's problem. That author thought the only rival to the hypothesis of even chance was one of providential control. More modestly, imagine for the sake of illustration that on prior grounds there is only one alternative to $h$, namely $i$, Nicholas Bernouilli's hypothesis about the distribution of male and female years, with the chance of a male year being $18/35$. What, according to the conceptions of Arbuthnot and Laplace, would be a wise rejection class for $h$? Apparently the class should include outcomes which would occur rarely, if $h$ were true, and rather less rarely, if $i$ were true. That is, the choice of $R$ should not depend on the likelihood of $\langle R, h \rangle$ alone, but on a

† P. S. de Laplace, *Essai Philosophique sur les Probabilités* (Paris, 1812). Translation of Truscott and Emory (New York, 1902).

comparison between the likelihoods of $\langle R,h \rangle$ and $\langle R,i \rangle$. Do not reject $h$ if what happens would happen rarely, if $h$ were true. This seems common-sense. Reject $h$ only if there is something better.

This kind of conclusion perfectly accords with our analysis of support in terms of likelihood ratios. So far in discussing tests I have examined Arbuthnot's problem from within his own conceptions, and even then we have been led to consider not likelihoods but likelihood ratios. But now approaching from a different direction, one might say simply that a test should reject an hypothesis only if some other hypothesis is much better supported than it. Relative support has been analysed in terms of relative likelihoods, and the conclusion is just as before: reject in terms of likelihood ratios.

However mere concurrence of different approaches cannot be held to settle that they are correct. What is required of a theory is that it should have plenty of consequences which are probably true and none which are probably false. This requires more precise statement of a theory of testing, which will not be given until the next chapter. But at once some favourable conclusions can be drawn. We have noted Arbuthnot might equally well have appraised his hypothesis in terms of the actual data derived from the registers, or in terms of a mere abstract of the data, simply mentioning the number of male years, and not their order. Why? This is a riddle which must be explained before any theory of testing can be accepted.

The riddle about Arbuthnot's problem is easily solved. Here consider only the special case of one rival hypothesis to $h$, say $i$. Both assume independent trials. Let $e$ be the registrar's data or any other data stating not only the proportion of male years, but also the actual order of male and female years. Let $d$ be a contraction of that data, stating only the proportion of male years among the 82. Then it is possible to prove, in general, that,

$$\frac{\text{Chance of getting } e, \text{ if } h \text{ is true}}{\text{Chance of getting } e, \text{ if } i \text{ is true}} = \frac{\text{Chance of getting } d, \text{ if } h}{\text{Chance of getting } d, \text{ if } i}.$$

That is, the ratio between the likelihoods of $h$ and $i$, on $e$, is exactly the same as the ratio of their likelihoods on the shorter data $d$.†

† R. A. Fisher, 'On the mathematical foundations of theoretical statistics', *Philosophical Transactions of the Royal Society*, A, CCXXII (1922), 309–68.

Fisher, who first studied the matter, called $d$ a *sufficient statistic* for $e$. Any information about the likelihood ratio which is conveyed by $e$ is also conveyed by $d$. This fact shows that the order and arrangement of male and female years is irrelevant to comparing the support for $h$ and $i$, or for comparing any other binomial hypothesis. Assuming independence, the order is not irrelevant to the likelihood, as Arbuthnot may have thought, but it is irrelevant to the likelihood ratio.

*Fisher's contention*

Explicit consideration of rival hypotheses within the theory of testing is due to Neyman and Pearson. Neyman first stated the principle that there could be no statistical test of an hypothesis without reference to rival hypotheses. But this principle has met many denials from Fisher. There is a forceful example in his final work on foundations.† He considers the hypothesis, first studied statistically in 1769, that the major stars are distributed randomly on the celestial sphere, that is, that the chance of a star being in any given area should be proportional to the area on the sphere. It is known that 5 stars are very near to the star Maia. He computes the chance, on the hypothesis of random distribution, of 6 stars falling this close together, and gets 1/33,000, which 'is amply low enough to exclude at a high level of significance any theory involving a random distribution'.

'The force with which such a conclusion is supported', continues Fisher, 'is logically that of the simple disjunction: Either an exceptionally rare chance has occurred, *or* the theory of random distribution is not true.'

This disjunction seems devoid of force. Even if the theory of random distribution were true, an exceptionally rare chance would have to have occurred. This is the old point about Arbuthnot's problem. We need, for instance, only rule the sphere into sufficiently many equidistant lines of longitude with one pole at, say, Arcturus, and a zero line running through Maia. Then however the stars be arranged on the sphere, the chance that stars should fall between the lines in any given way will be very small, much much less than 1/33,000, as small as we like and care to rule lines.

† R. A. Fisher, *Statistical Methods and Scientific Inference* (London, 1956), especially pp. 37 ff.

Hence, if Fisher's disjunction had any force, we should always have to exclude any hypothesis like that of random distribution, whatever happened. So it has no force.

I think that Fisher never tells us, in all his manifold writings, much more about the logical force of tests. Take a test he values, and whose explications and ubiquitous use in practice are largely due to himself. This is Gossett's $t$-test, used for testing the hypothesis that a Normal distribution has mean $\mu$. For given statistical data $d$ Gossett computes a function $t(d,\mu)$. The distribution of $t$ for any value of $\mu$ is known. Hence it is possible to compute regions of outcomes, based on $t$, with very low chance of occurring. This provides the celebrated test. Gossett, says Fisher, 'was usually content to leave the inference in the form of a test of significance, namely that, on the hypothesis, for example, that the true mean of the population was zero, the value of $t$ observed, or a greater value, would have occurred in less than 1 % of trials, and was therefore significant at, for example, the 1 % level'.†

At no time does Fisher state why one is allowed to add the clause 'or a greater value' so as to form a region of rejection. Nor does he state fully why just this contraction of the data, from $d$ to $t(d,\mu)$, is such a good one. It is possible to prove that the $t$-test is excellent. Fisher himself showed that, *if the possible hypotheses are all Normal distributions*, $t$ is a sufficient statistic, that is, as far as relative likelihoods are concerned, conveys as much information as the original data. For a *certain class* of rival hypotheses, just this abstraction is correct.

In his polemical moods Fisher has strongly criticized recent theorists of statistics for their ignorance of experimental techniques, opposing them to the men who first felt the need for work like the $t$-tests, 'who first conceived them, or later made them mathematically precise' and who were, notably, 'all actively concerned with researches in the natural sciences'. He suspects recent theories would never have been propounded if their authors 'had any real familiarity with work in the natural sciences, or consciousness of those features of an observational record which permit of an improved scientific understanding, such as are particularly in view in the design of experiments'.‡ One of several butts of this

† Pp. 80 f.            ‡ Pp. 75 f.

remark seems to be the use of rival hypotheses in framing a theory of testing.

Hence it may be salutory to conclude with what W. S. Gossett said eighteen years after he invented the $t$-test. Gossett was a practical statistician working for Guinness, the brewers; he combined a piercing appreciation of abstract theory with a firm grasp of experimental conditions. His common-sense is a refreshing antidote to many an obscure theory. He writes that a test

doesn't in itself necessarily prove that the sample is not drawn randomly from the population even if the chance is very small, say ·00001: what it does is to show that if there is any alternative hypothesis which will explain the occurrence of the sample with a more reasonable probability, say ·05 (such as that it belongs to a different population or that the sample wasn't random or whatever will do the trick) you will be very much more inclined to consider that the original hypothesis is not true.†

Hence even Fisher's arguments *ad hominem* are devoid of force. Here is a man 'actively concerned with researches in the natural sciences', who 'first conceived' one of the great tests, and who urges that it is not merely low likelihood which matters, but rather the ratio of the likelihoods.

*Hypotheses and data*

The next chapter will explain two theories of testing which assume that the best test of an hypothesis depends not only on the hypothesis under test, but also on rivals to it. But how know the class of possible rivals? The problem is not important for the theory, for tests can be defined for any hypothesis, against any possible class of rivals. But it is crucial in practice, if only to exclude computations based on a host of different sets of rivals.

First a mere matter of terminology. In the next chapter testing will be discussed more formally within the theory of support. There the class of possible distributions will be included in what I call the *statistical data*, which is expressed by a joint proposition. A joint proposition is simply one which states, jointly, the class of possible distributions and a possible outcome for a definite trial on the set-up. This method of description is not the usual one, for statisticians customarily refer to the result of particular experiments

† From a letter printed in E. S. Pearson's appreciation of Gossett; *Biometrika*, XXX (1938), 243.

as data, and prefer to call the class of possible distributions the *specification* of the problem. The picture is of a mathematician who designs a test fitting a certain specification, and which will suit all future obtained data. However, I lump both specification and experimental results under the single head of statistical data, because, from the point of view of the experimenter, these provide the datum line from which he works. Data, as the *Dictionary* says, is what is known or assumed for purposes of further calculation. Testing is relative to data; in calling it data I do not mean to canonize it into infallibility, but only to indicate its place in the relation.

Perhaps the corrigibility of what I call data must be emphasized. Everything is fallible. My statistical data cover both experimental results and a class of possible distributions. It is possible to be wrong about an experimental result. In any long experiment it is, indeed, almost certain that some of the readings will be wild, either through mechanical failure or clerical ineptitude. Distinguishing wild observations on statistical grounds has been one of the oldest of statistical problems, and has received new theoretical impetus from recent work.† In the same way a claim about a class of possible hypotheses is equally open to test against some wider class of rivals. There is nothing sacred about data.

In a particular testing situation some joint proposition may serve as statistical data. How is it arrived at? This varies from case to case. Sometimes it is easy. Often one can tell by examining the mechanism of an experiment, and its resemblance to other well-known experiments, that outcomes of one trial are independent of any other. The statistical hypothesis of independence can itself be tested later. Hence one excludes, for the time being, rival hypotheses assuming dependence. Often this may be done deliberately: the set-up may be designed so that outcomes are independent. This is accomplished by randomization, in which a well-known, independent set-up (perhaps based on tossing a coin or looking at specially prepared tables of random sampling numbers) is used in picking out individual outcomes from the set-up under test.

Again it may be known from other experiments that the distribution of chances in some set-up will belong to a definite family

† J. W. Tukey, 'The future of data analysis', *Annals of Mathematical Statistics*, xxxiii (1962), 1–67.

of distributions. It has been discovered that the distribution of chances of emission of a radioactive particle in a certain time interval, and the distribution of chances of waiting times at a telephone exchange, and, as the text-books remind one, chances that in any given year a Prussian soldier will have been kicked to death by his mule, all are connected with a certain family of distributions, named after Poisson, who investigated it in the 1820's. Members of such families can often be uniquely specified by one or two parameters, and the design of tests may be relatively easy.

A theoretical point deserves mention. The problem of statistical inference, as stated in the opening sentence of this essay and slightly refined later, is to give a set of principles which validate those correct inferences which are peculiarly statistical. This means to give a set of principles validating correct inferences in which terms like 'chance', 'outcome' and so forth do occur essentially. A term occurs essentially in an inference if replacing it throughout the pattern of inference by another term of the same grammatical type can produce a fallacious inference. Now I think that inferences to what is best supported by statistical data, or to what can be rejected on the basis of statistical data, are indeed peculiar to statistics. But inferences to statistical data do not, in general, seem to me to be peculiar to statistics at all.

To take the easiest case, suppose you are interested in the radioactive decay of a new substance and take as datum that the possible distributions are of the form $F(t) = 1 - \exp(-\lambda t)$. Where does this come from? Three considerations: simplicity, analogy, and general physical theory. Simplicity, because the stated form is mathematically tractable. Analogy because similar set-ups, using different radioactive substances, are believed to display a distribution of that form. General theory, because a theory about radiation predicts a distribution of that form. These are the lynchpins of the inference. Whether or not the principles of such inference can be stated I do not know. I hope so. But it is plain that its principles are not peculiar to statistics: exactly the same form of argument might be given in quite non-statistical contexts. So it cannot be part of the foundations of statistics to provide principles validating such inference to statistical data. And at the moment we are in search of the foundations of statistics, not of a whole philosophy of science.

Radiation is perhaps too simple a case because there is so clear-cut a general law. In other matters simplicity and analogy are the most germane desiderata. In every case, whether it involves the use of clear-cut general theory or rests entirely on empirical guess-work, the statistical data represent a more or less entrenched conjecture, a convenient base for peculiarly statistical considerations, but which is always subject to future rejection.

Much ingenuity goes into framing statistical data. In lieu of general theory one must often merely construct a hopeful model of the experimental situation embodying as much as one knows or dares conjecture. In statistical writing, the model is sometimes taken as being something different from data; it is part of the specification of a problem. But a model is only a conjecture. A good example is the urn model for the epidemic, described on p. 9 above. The model suggests a family of possible distributions for describing an epidemic. Once discovered, members of this family may be specified by three parameters (corresponding to numbers of black and red balls originally in the urn, and the number of red balls to be added after drawing a red ball). But the hypothesis that the true distribution belongs to this family is itself statistical, and may be tested against other rival hypotheses, perhaps themselves suggested by urn models.

*Changing chances*

Mention of the urn model for an epidemic recalls from the first chapter one of our requirements for a good analysis of long run frequency. Not only should our analysis obviate reference to the 'long run', but also it should provide for changing chances in which no long run at constant chance is even possible. All statisticians and probability theorists are content to work with such ideas, though it is unclear how this can be permitted if they are really talking about long runs as they say. In fact long runs are otiose. A theory of testing in terms of relative likelihoods makes plain, for instance, that to test an hypothesis one need never suppose, as part of the theory, that more than one instance of any kind of trial has ever occurred.

Imagine a laboratory worker wishing to test one hypothesis about an experiment against another. He thinks that experiments of some sort are independent of each other; he calls them trials of kind $K$ and makes 100 such trials. But in testing on the basis of this data we do not need to refer to 100 trials of some kind and to hypotheses about them; we do not have to pretend that we know that there are 100 repetitions of an independent experiment. The hypothesis under test assumes independence, but not the theory of testing, which considers the relative likelihood of two joint propositions. The joint propositions are not directly about trials of kind $K$, but about a single compound trial of kind $K'$ say, consisting of 100 experiments, independent or not. Test hypotheses about trials of kind $K'$, by comparing the relative likelihoods. There is never any need to imagine that more than one trial of kind $K'$ has been performed or could be performed.

Hence there is no difficulty in testing hypotheses about changing chances in the epidemic. A trial of kind $J$, say, may be conceived of as over the whole length of the epidemic, and one need not pretend that just this epidemic could occur again, let alone a long run of such epidemics. The chance of 40 new infections on day 4 of the epidemic is analysed as the chance of a certain outcome on a trial of kind $J$, namely the outcome for which anything at all may happen on days 1–3, and days 5 plus, and on which there are 40 new infections on day 4. This gives good sense to talk of the chance of infection changing from day to day. Similarly hypotheses about the distribution of chances of infection may be tested. If the experimental data consist of the numbers of new infections on days 10–21, then one tests hypotheses about the distribution on trials of kind $J$, using as datum that the result of the known trial of kind $J$ consisted of any possible numbers of infections except for days 10–21, when the numbers were as stated. Nowhere does one need to speak of a long run of days 10–21.

*Kinds of trial*

An increasing battery of terms is being introduced, and their interrelations are being defined with some precision: joint proposition, statistical data, likelihood, rejection, support, chance, outcome, kind of trial. This is traditionally the way to gain analytic mastery of a complex discipline. But it is no part of my thesis that the application of these terms will be entirely simple in every practical situation. Take the useful idea of a kind of trial. In the preceding section it is argued that in testing an hypothesis one need never assume, as part of the theory of testing, that more than one trial of some given kind has ever occurred, although, of course, the hypothesis under test may make this assumption. But there is still a problem in suitably describing a trial of some particular kind.

Suppose you are concerned with hypotheses about the frequency with which chrysanthemums of some variety give purple flowers. You make an experiment which consists in planting ten chrysanthemums; your plan is to note the proportion of these ten plants which flower purple. But you step on one of the seeds and so it fails to germinate. What kind of trial have you made? A trial which consists in planting nine successful seeds? Or ten, of which one fails to germinate either through insects, infection or frost or negligence or whatever? Or a trial in which ten are planted and one is stepped on? And what are the possible results of the trial you have made?

The author of this example has suggested a way to analyse the experiment.† But it is unclear, as I think he would agree, that there is any definitively correct analysis. Yet this is the sort of thing with which a field worker is continually confronted. The psychologist discreetly forgets the subject who fell ill half-way through his memory test. The chemist may ignore the experiment which he never completed because he was summonsed by a telegram. What kinds of trials have actually been made? In truth one will usually answer in the way which makes the analysis most convenient. This is a matter which can be too easily forgotten in philosophizing about science. Its existence should not impede precise analysis, but it should stop undue inflation of that precision.

† G. A. Barnard, 'The meaning of a significance level', *Biometrika*, XXXIV (1947), 179–82.

CHAPTER VII

# THEORIES OF TESTING

Henceforth it will be assumed as proven that an adequate theory of testing must consider not only the statistical hypothesis under test, but also rivals to it. This may be common-sense: 'don't reject something unless you've something better'. It involves a conception now becoming general in the philosophy of science, and which is currently striving to oust the former idea that an hypothesis could be rejected independently of what other theories are available.

Even if this kind of attitude be agreeable, it is far from evident how rival hypotheses are to be weighed. There may be, as has already been warned in the last chapter, radically different kinds of test. The sequel will show that controversy between classical theories of testing stems in part from their trying to answer different questions. The most famous theory, due to Neyman and Pearson, will be shown to have very limited application. For the present I shall continue to regard the central problem as that of tests which can imply, on the basis of experimental results, whether or not an individual hypothesis should be rejected. Tests thus divide possible experimental results into two classes, those which suffice to reject the hypothesis, and those which do not. The former class of results will be called the *rejection class*.

*Likelihood tests*

Our theory of support leads directly to the theory of testing suggested in the last chapter. An hypothesis should be rejected if and only if there is some rival hypothesis much better supported than it is. Support has already been analysed in terms of ratios of likelihoods. But what shall serve as 'much better supported'? For the present I leave this in abeyance, and speak merely of tests of different stringency. With each test will be associated a critical ratio. The greater the critical ratio, the more stringent the test. Roughly speaking hypothesis $h$ will be rejected in favour of rival $i$ at critical level $\alpha$, just if the likelihood ratio of $i$ to $h$ exceeds $\alpha$.

Hence suppose we have data stating that the true distribution on trials of some kind is a member of the class $\Delta$. By a simple statistical hypothesis we mean one of the form, 'The true distribution is $D$', where $D$ will be in $\Delta$. A complex hypothesis is of the form 'The true distribution is in $\Gamma$' where $\Gamma$ is contained in $\Delta$. What will be the rejection class for rejecting first of all, a simple hypothesis, and then, a complex hypothesis?

It is easy for simple hypotheses. Simple hypothesis $h$ will be rejected in the light of result $E$ just if there is some $i$, a simple hypothesis consistent with our data, such that

$$\frac{\text{Chance of getting } E \text{ if } h \text{ is true}}{\text{Chance of getting } E \text{ if } i \text{ is true}} < \alpha. \qquad (1)$$

So the rejection class $R$ consists of all possible results for which the above inequality holds for some $i$.

It is much the same for a complex hypothesis. Let us say the simple hypothesis 'The true distribution is $D$' is *a member of* the complex hypothesis, 'The true distribution is in the class $\Gamma$', just if $D$ is a member of $\Gamma$. Then the complex hypothesis $H$ shall be rejected if a result $E$ occurs such that for every simple hypothesis $h$ in $H$, there is some $i$, consistent with the statistical data, for which the inequality (1) holds. The rejection class for $H$ is the class of all such possible results. If, according to the statistical data, the hypothesis $J$ must be true if $H$ is false, we shall sometimes say that this is the rejection class for testing $H$ against the rival $J$.

This rejection class exemplifies what I shall call a *likelihood test*. A more extensive definition is to hand in terms of joint propositions. To recall the terminology, a simple joint proposition $\langle X, K, D; T, K, E \rangle$ states that the distribution of chances on trials of kind $K$ on set-up $X$ is $D$, and that $E$ occurs on trial $T$ of kind $K$. A complex joint proposition, or for short a joint proposition, $\langle X, K, \Delta; T, K', E \rangle$, states that the distribution on trials of kind $K$ is a member of the class $\Delta$, and that $E$ occurs on trial $T$ of kind $K'$. A simple joint proposition $\langle X, K, D; T, K, E \rangle$ is included in a joint proposition $e$ just if $e$ is equivalent to the joint proposition $\langle X, K, \Delta; T, K, E' \rangle$, and $D$ is in $\Delta$, while $E$ is contained in $E'$.

Then we say that a simple joint proposition $h$ is *$\alpha$-rejectable* by $e$ if either (i) $h$ is inconsistent with $e$, or (ii) there is a simple joint proposition $i$, included in $e$, such that the likelihood ratio of $i$ to $h$ exceeds $\alpha$.

A likelihood test with critical ratio $\alpha$ is, essentially, one in which an hypothesis is rejected at level $\alpha$ if it is $\alpha$-rejectable by $e$. There is only one minor qualification to ensure complete generality. Moreover, a test will naturally reject an hypothesis which is equivalent to a whole class of hypotheses which it rejects—if $h$ is true if and only if some member of the class is true, and every member of the class is to be rejected, then $h$ too must be rejected.

*A likelihood test with critical ratio $\alpha$ is one in which $h$ is rejected by data $e$ if either (i) $h$ is $\alpha$-rejectable by $e$, while $e$ & $h$ together do not entail any simple joint proposition not $\alpha$-rejectable by $e$, or (ii) $h$ is equivalent to a class of propositions each answering to condition (i).*

It may be checked that the special cases with which this section began are indeed likelihood tests under the above definition: our definition prescribes rejection classes such as were defined earlier. Now it will be useful to contrast this new theory of testing with others more familiar from the statistical literature.

*Likelihood ratio tests*

Despite superficial similarity likelihood tests are not the same as what are called *likelihood ratio tests*. An early paper of Neyman and Pearson explained this different notion at a powerful and theoretical level.† It was a predecessor of their later work, which we shall examine presently. I mention the earlier likelihood ratio theory only in order to distinguish it from our likelihood theory.

When a simple hypothesis $h$ is being tested against a simple rival $i$ in the likelihood ratio theory, we choose a rejection class $R$ satisfying the following condition:

$$\frac{\text{Chance of getting } R \text{ if } h \text{ is true}}{\text{Maximum (chance of } R \text{ if } h \text{ is true, chance of } R \text{ if } i \text{ is true)}} < \alpha,$$

where $\alpha$ is some constant. This does not define a unique $R$: the best $R$ satisfying this condition is chosen by yet another criterion. But for given $R$, $h$ is rejected if the observed result is a member of $R$. Though this kind of test sometimes works out to be a likelihood

† J. Neyman and E. S. Pearson, 'On the use and interpretation of certain test criteria for the purposes of statistical inference', *Biometrika*, XXA (1928), 175–240, 263–94.

test, it need not. The reason is, essentially, that in likelihood tests we demand that individual results satisfy a certain condition, and form the rejection class $R$ of those results; in the likelihood ratio theory, we demand that a class of results $R$ satisfies a certain condition. The two conditions may coincide but in general do not. It is interesting that several likelihood ratio tests are agreed by everyone to be absurd.[†] Not one of these is a likelihood test. So we shall not consider likelihood ratio tests again; when I speak of likelihood ratios I shall merely mean ratios of likelihoods, and not any concept peculiar to the likelihood ratio theory of testing.

*The Neyman–Pearson theory*

The mature theory of Neyman and Pearson is very nearly the received theory on testing statistical hypotheses.[‡] It has been made increasingly sophisticated in recent years, but the central ideas have not been altered. In explaining, let us first forget about likelihood ratios altogether.

According to this theory, there should be very little chance of mistakenly rejecting a true hypothesis. Thus, if $R$ is the rejection class, the chance of observing a result in $R$, if the hypothesis under test is true, should be as small as possible. This chance is called the *size* of the test; the size used to be called the significance level of the test.

In addition to small size, says the theory, there should be a good chance of rejecting false hypotheses. Suppose simple $h$ is being tested against simple $i$. Then, for given size, the test should be so designed that the chance of rejecting $h$, if $i$ is true, should be as great as possible. This chance is called the *power* of $h$ against $i$.

Thus, in this case of what has been called simple dichotomy, the Neyman–Pearson theory instructs one to choose a test with small size and great power. There is in general no way of jointly minimizing size and maximizing power. Neyman and Pearson originally recommended that one choose a small size, say 0·01, and then

[†] E. L. Lehmann, 'Some principles of the theory of testing hypotheses', *Annals of Mathematical Statistics*, XXI (1950), 2; *Testing Statistical Hypotheses* (New York, 1959), pp. 252 f. M. G. Kendall and A. Stuart, *The Advanced Theory of Statistics* (3 vol. ed.) (London, 1961), II, pp. 246 f.

[‡] J. Neyman and E. S. Pearson, 'On the problem of the most efficient tests of statistical hypotheses', *Philosophical Transactions of the Royal Society*, A, CCXXXI (1933), 289–337.

maximize the power consistent with that size. Recent workers have constructed more elaborate theories for choosing size and power in the light of many factors, including even the cost of experimentation. Under certain conditions rather large sizes have been permitted.

A slight generalization enhances the elegance of some facts about Neyman–Pearson tests. We shall hardly use this generalization, but it is not to be ignored in forming an appreciation of the theory. A *pure* test will be one of the sort we have already described, in which possible results are divided into two classes, one of which is the rejection class. A *mixed* test employs a third possibility. Let there be another quite unrelated chance set-up in which the distribution of chances is well known; it might involve tossing a coin or using a table of random numbers. In a mixed test, results are divided into three categories; those in which the hypothesis is rejected, those in which it is not, and those in which whether or not to reject depends on the outcome of some unrelated experiment, like tossing a coin. Thus a mixed test consists not so much of a rejection class as of a rejection function, $r(E)$; if $r(E) = p$, then if event $E$ occurs, the hypothesis is to be rejected if and only if on a trial on a table of random numbers, some designated event with chance $p$ also occurs. In the case of a pure test, for any given $E$, $r(E)$ is, of course, either 0 or 1.

The *Fundamental Lemma of Neyman and Pearson* states that in the case of simple dichotomy, there exists, for any possible size, a uniquely most powerful test of that size. It may be a mixed test, but it is unique. The proof of the lemma implies that this unique test is also a likelihood test (so long as we allow for mixed likelihood tests) and also that for simple dichotomy every likelihood test is a most powerful test of some size. Hence the case of simple dichotomy does not distinguish between the theory of Neyman and Pearson and the theory of likelihood tests.

*Complex hypotheses*

The size of a test $R$ of simple hypothesis $h$ was defined as the chance of getting $R$, if $h$ is true. The size of a complex hypothesis $H$ is conveniently defined as the maximum of the sizes of $R$ when used as a test of any simple hypothesis in $H$. Now suppose complex hypothesis $H$ is being tested against complex $J$. It may happen that a single test of given size is most powerful for every simple hypothesis in $H$ against every simple hypothesis in $J$. If so, this is the *uniformly most powerful test*. Under very general conditions, UMP tests, if they exist, are likelihood tests, but likelihood tests are not necessarily UMP tests. In fact, though UMP tests do exist for simple dichotomy, as proved by the fundamental lemma, they seldom exist elsewhere. So other desiderata must be imported for defining optimum tests. These do not always lead to likelihood tests, and hence it is possible to distinguish our two theories of testing. First, however, we must scrutinize the rationale of the Neyman–Pearson theory, and especially these striking concepts, size and power.

*Size and power*

At first sight the Neyman–Pearson theory entirely accords with common preconceptions about what a good test should be like. 'A small chance of rejecting what is true, combined with a large chance of rejecting what is false.' But closer inspection shows this charm to be illusory. Chances—long run frequencies—do not exist in a vacuum: a chance is a propensity to have a certain long run frequency on trials of some definite sort on some distinct chance set-up. When speaking of chances, the kind of trial must be specified.

The chance of rejecting $h$ if it is true turns out to be the chance of getting some result on a trial of kind $K$ on a given set-up, such that, according to the test we have in mind, $h$ is to be rejected. Let us suppose this chance is small, say 0·01. What is implied?

First, the chance that you will get a result which leads you to reject $h$, is indeed small, being exactly 0·01. Hence, for instance, *lacking data about the result of a trial of kind $K$*, the hypothesis that you will not mistakenly reject $H$ is very well supported. Hence if you must speculate on mistaken rejection according to some test *before* a trial has been made on the set-up, the size of the test is crucial. But even if before you know the result of the trial of kind $K$, there is good support for the hypothesis that, using the test in question, you will not mistakenly reject $h$, it does not follow that there is good support after the result is known. A trivial example will prove this.

Let there be a set-up $X$, trials of kind $K$, and three possible results of such trials, $E_1$, $E_2$, and $E_3$. Let there be only two statistical hypotheses, $h$ and $i$, which assign chances as shown on the table below. One test of size 0·01 would reject $h$ for $i$ if and only if $E_1$ occurs. Before making a trial of kind $K$, one can be pretty confident that this test will not lead to mistaken rejection of $h$. The chance is only 0·01. But if the unusual event $E_1$ does occur, we can be certain that, on this occasion, the test led us astray. For occurrence of $E_1$ proves conclusively that $i$ is not true, and hence that its rival, $h$, is true. Similarly if the rival hypotheses are $h$ and $j$; although occurrence of $E_1$ does not prove $h$ is true, it gives very good reason to suppose it is true and $j$ is false.

|   | $P(E_1)$ | $P(E_2)$ | $P(E_3)$ |
|---|---|---|---|
| $h$ | 0·01 | 0·95 | 0·04 |
| $i$ | 0 | 0·95 | 0·05 |
| $j$ | 0·00001 | 0·95 | 0·04999 |

Of course the silly test just described is not the most powerful test of $h$ against the rival, and it may be retorted that the Neyman–Pearson theory would never countenance such a test. My present point is only this: at first sight small size—small long run frequency of mistaken rejection—seems to mark a test out as intrinsically desirable. And indeed if you were concerned with a unique hypothesis and had to bet, before a trial had been made on the relevant set-up, whether or not a test would mistakenly reject the hypothesis, you would be interested in the size and nothing but the size. All tests of the same size would furnish as good a bet. But after making a trial on the set-up, not all tests of the same size would furnish equally good bets. In the above case, if $E_1$ were observed, rejection based on the test just stated would be disastrous.

Now let us apply these elementary observations to power. Suppose we were to place *before-trial bets* on tests of an hypothesis, subject to the condition (*a*) if the hypothesis is true and we reject it, we lose our stake, but if true and we do not reject, we win a prize, and (*b*) if the hypothesis is false and we reject it, we win a prize, and lose some sum if we do not reject. The bet is to be decided on the basis of our rejection in the light of a trial on the set-up. But we must choose our test before making a trial. Then size and power would be of prime interest. But what of *after-trial bets*, or evaluation of the hypothesis in the light of the result yielded by an actual trial? It is worth citing two tests, $R$ and $S$, such that $R$ and $S$ have the same size, and $R$, being more powerful, is the best test for before-trial betting and yet such that, after a trial, it is possible that $S$ should be preferable.

|   | $P(E_1)$ | $P(E_2)$ | $P(E_3)$ | $P(E_4)$ |
|---|---|---|---|---|
| $h$ | 0 | 0·01 | 0·01 | 0·98 |
| $i$ | 0·01 | 0·01 | 0·97 | 0·01 |

Once again suppose a set-up and kind of trial, and that there are two hypotheses; this time suppose there are four possible results. Chances are shown above. Define test $R$ as rejecting $h$ if and only if $E_3$ occurs, so that its size is 0·01 and its power 0·97. Test $S$ will be such as to reject $h$ if and only if $E_1$ or $E_2$ occur; its power is a mere 0·02. Hence subject to conditions (*a*) and (*b*), $S$ is much inferior to $R$ for before-trial betting. But, if $E_1$ is afterwards observed, then $S$ is on this occasion much to be preferred to $R$, for $S$ would reject $h$, which is certainly false, while $R$ would not reject it. Thus a test which is preferable for before trial betting need not be preferable after the trial has been made.

There is nothing paradoxical about this situation, nor is it peculiar to statistics. Suppose you are convinced on the probabilities of different horses winning a race. Then you will choose among bookmakers he who gives you what, in your eyes, are the most favourable odds. But after the race, seeing that the unexpected did in fact happen, and the lame horse won, you may wish you had wagered with another bookmaker. Now in racing we must bet before the race. But in testing hypotheses, the whole point of testing is usually to evaluate the hypotheses after a trial on the set-up has been made. And although size and power may be the criteria for before-trial betting, they need not be the criteria for after-trial evaluation.

Yet these considerations do not refute the Neyman–Pearson theory as a theory of after-trial evaluation. For it is not a theory just about size and power; it says, for instance, that the uniformly most powerful test, if it exists, is the best. For simple dichotomy the uniformly most powerful test is a likelihood test. Likelihood tests are not open to the counterexamples just given. Thus when UMP tests exist, the Neyman–Pearson theory does give an appropriate account of testing: not because low size and great power are intrinsically desirable for after-trial evaluation, but because they entail a genuine desideratum. But now we shall establish that when no UMP test exists, what the Neyman–Pearson theory claims as best is not necessarily desirable.

*Unbiased and invariant tests*

In order to test the complex hypothesis $H$ against the complex hypothesis $J$, a trial is to be made on a set-up. $H$ will be rejected if and only if the result of the trial is in the rejection class $R$. To many thinkers it has seemed absurd to use an $R$ such that the chance of $R$, on some member of $H$, exceeds the chance of $R$, on some member of $J$. Neyman and Pearson agree, and hold that the power of a test should exceed its size. A test meeting this requirement is called *unbiased*. Sometimes even when no test is strictly UMP, one possible test is uniformly most powerful among unbiased tests. Such a UMPU test, as it is called, is favoured by Neyman and Pearson.

Failing such a test, they use yet another consideration. No test should depend on the way in which the hypotheses are named: tests should be invariant under different naming procedures. Or to put it differently, if the hypotheses under test are in some way symmetric, the test should display the same symmetry. Sometimes there is an unbiased test which is uniformly most powerful among invariant tests and the Neyman–Pearson theory advises use of such a UMPI test. Now we shall exhibit a UMPI test which is *not* a likelihood test, and so discriminate between the two theories of testing.

Let set-up $X$ be subject to trials of the following kind: a trial may have 101 possible results, labelled from 0 to 100. There are also, let us suppose, 101 possible hypotheses about the set-up. $H$ is the hypothesis under test; according to it the chance of result number 0 shall be 0·90, and of every other result, 0·001. The rival hypothesis $J$ includes 100 simple hypotheses $J_1, J_2, ..., J_{100}$. According to $J_n$ the chance of getting result 0 is 0·91, and of getting the result numbered $n$ is 0·09; the chance of getting a result not numbered 0 or $n$ is, according to $J_n$, absolutely zero. Thus the chances are as on the table below. The problem is to find a test of size 0·1 based on the result of a single trial on this set-up.

Intuitively speaking, if the outcome is 0, it scarcely discriminates between $H$ and $J$: 0 is about as likely to occur if $H$ is true as if it is false. But if a result other than 0 occurs, the case is very different. If 3 occurs, we can at once rule out every member of $J$ except $J_3$. And if $J_3$ is true, something fairly common has occurred, while if $H$ is true, something which happens only one time in a thousand has occurred. So occurrence of 3 might give some indication that $H$ is false and $J$ true. And it happens that the likelihood test of size 0·1 rejects $H$ if and only if an event $n > 0$ occurs.

|  | $P(0)$ | $P(n)$ | $P(m)$ for $m \neq n$ |
|---|---|---|---|
| $H$ | 0·90 | 0·001 | 0·001 |
| $J_n$ | 0·91 | 0·09 | 0 |
|  | $1 \leqslant n \leqslant 100$ | $1 \leqslant m \leqslant 100$ |  |

But this test is not unbiased; its size is 0·1 but its power is 0·09, which is less than the size. There is, however, a test which is uniformly most powerful among invariant tests. It turns out to be a mixed test. We take some auxiliary unrelated set-up with two possible results, $A$ and $B$, the chance of $B$ being 1/9. Call this the randomizer. Now consider the test which rejects $H$ if and only if 0 does occur, and the randomizer has result $B$. The size of this test is once again 0·1; its power is a little more than 0·1. So the test is unbiased, and it is uniformly most powerful among invariant tests.

The likelihood test says: reject $H$ if and only if 0 does *not* occur. The UMPI test says: reject $H$ if and only if 0 *does* occur, and the randomizer has result $B$. No better contradiction could be hoped for: these tests never agree to reject $H$.

If one were thinking of using the UMPI test for evaluating $H$ in the light of the result of a trial, it would be very odd. We would get a kind of paradox. Suppose 3 occurred. Then the only possible hypotheses are $H$ and $J_3$. And the UMPI test of $H$ against $J_3$— indeed, the UMP test— would reject $H$ for $J$. Yet we find that the UMPI test for $H$ against $J$ tells us to reject $J$ for $H$ in this very circumstance, so long as a quite unrelated event also occurs.

This looks like a contradiction in the Neyman–Pearson theory. The air of paradox is relieved by considering what size and power mean. If one were told to select some unbiased test before making any trial on the set-up, and (*a*) if $H$ is true, win $r$ if your test does not reject $H$, and lose $s$ if it does, but (*b*) if $J$ is true, win $r'$ if your test does reject $H$, and lose $s'$ if it does not, then you would be wise to choose the UMPI test above. The UMPI test is the test for *before-trial betting* in the conditions given. Mind you, these conditions are pretty esoteric, but perhaps something similar could arise in real life. However, if you were asked to evaluate $H$, according to the same conditions, *after* you had made a trial on the test and had observed 3, it would no longer be wise to employ the UMPI test.

Thus we infer that *the Neyman–Pearson theory is essentially the theory of before-trial betting*. It is not necessarily the theory for evaluating hypotheses after trials on the set-up in question have been made, and the results discovered. In all the cases studied so far the theory of likelihood tests does, however, seem appropriate to the latter situation.

## *'Worse than useless' tests*

The likelihood test in the situation just described is of a sort which, in the literature, has been called worse than useless. Its power is less than its size. Thus without making any trial at all on the set-up we can devise a test with greater power. With a table of random numbers, devise a set-up in which the chance of some event, say $A$, is 1/10. Make a trial on this set-up, and reject $H$ if the result is $A$. The size of this test is 0·1, and so is the power. This test is more powerful than our likelihood test. So by, in effect, tossing a coin one could have a more powerful test than the likelihood test. Hence such a test as the likelihood test has been called worse than useless.

But in fact calling the test useless is due to paying too much attention to size and power. If you had to choose a test for before-trial betting, then the likelihood test could truly be called worse than useless. You would be better to opt for entirely random rejections than for using that test. But if you are using the test for evaluating the hypothesis $H$ in the light of a result on the experiment, it would be foolish to use the random rejection: instead reject $H$ if and only if o does not occur.

## *The domain of the Neyman–Pearson theory*

The complete analysis, even of the Neyman–Pearson theory, is best located in the logic of support. I have been describing Neyman–Pearson tests as suitable for before-trial betting, but not for after-trial evaluation. The situation is more complex, and four possibilities must be distinguished. In each case we suppose a set-up $X$ is under investigation, and that we are interested in trials of kind $K$. A statistical hypothesis $H$ is to be tested. An initial body of data $d$ states the possible distribution of outcomes on trials of kind $K$; a test is to be based on the result of a single trial.

(1) *Before-trial betting.* Suppose I have in mind a Neyman–Pearson test of low size and great power. Before making any trial on the set-up I decide to use this test. I can compare two hypotheses, $h$ and $i$, which are as it were meta-hypotheses, namely hypotheses about how my test will treat the statistical hypothesis $H$. $h$ states that as a result of making my trial of kind $K$, my test will lead me to treat $H$ wrongly, rejecting it if true or retaining it if false. $i$ states that as a result of my trial my test will lead me to treat $H$ correctly. In virtue of the low size and great power, I can assert, solely in virtue of the law of likelihood, that *the initial information d supports i much better than h.* If before a trial is made $d$ constitutes all my data, then, in a before-trial betting situation, my choice is wise.

(2) *Complete after-trial evaluation.* I make a trial of kind $K$, observe the result $E$, and use a likelihood test to reject $H$. That is, *I assert that data d, and the fact that E occurred, support H much less well than some rival.* Since my data consist only of $d$ and the fact $E$ occurred, I justly reject $H$.

(3) *Incomplete after-trial evaluation.* I begin with the same Neyman–Pearson test as used in (1), with, we shall suppose, a rejection class $R$. I observe some result $E$ and note that it lies in $R$. In virtue of low size and great power, I can assert, *data d, and the fact that the result was in some (unspecified) member of R, support H much less well than some rival.* But I am not automatically entitled to reject $H$, for presumably I know more than that some unspecified member of $R$ occurred. I know that the specified member $E$ occurred. Our earlier examples have shown that in general we cannot be sure that this extra knowledge will leave the evaluation of $H$ unchanged. Indeed if we were testing one simple hypothesis against another, and used the UMP test, then in virtue of the Fundamental Lemma we would be sure that our test was a likelihood test, and hence could argue as in (2). But not in general. So in general we shall have to call the Neyman–Pearson method one for incomplete evaluation.

(4) *Economical after-trial evaluation.* But now suppose that before using the self-same Neyman–Pearson test I had decided before making the trial to observe merely whether the result lay in $R$ or not, and not to note the actual result. Or suppose I had decided to discard all information except that $R$ did, or did not, occur. Then just as is italicized in (3), I can assert that $d$ plus occurrence of $R$ supports $H$ much less well than a rival hypothesis. But unlike (3), $d$ and $R$ complete, by fiat, my available data. So now in virtue of my earlier decision, I am in a position to reject $H$.

But when can I justly discard or ignore data? Well, it may be that observation of individual results is costly, or that their analysis is tedious; at the same time, with low size and great power, it is a good before-trial bet that a test based merely on occurrence or non-occurrence of $R$ will not seriously mislead. In particular, if I am to use the same test on many hypotheses, say about successive items produced on an assembly line, and if I am concerned with keeping the average error as low as possible, and at the same time want a cheap test, then I may deliberately ignore or discard available data. One of the most ancient uses of such reasoning is in the custom of life insurance. An insurer can get a wealth of unanalysable data about anybody, but usually he does not even try. A 'trial' of a prospective client consists of having him medically examined. But it is a good before-trial bet that one's assessment of his chances will not be far wrong if based on the very limited data obtained by a half-hearted company physician. So only these data are obtained, and premiums are computed on their basis. A great deal of nonsense has been written about the logic of an insurer, but it comes down to this: you discard most of the data as a before-trial bet, and with what is left you compare the support for hypotheses about a man's life.

We see that *if accompanied by a rational decision to discard or ignore data, the Neyman–Pearson theory may be perfectly suitable for after-trial evaluation.* But when so used it is not an alternative to likelihood testing, but only a special case of it. We reject $H$ if $R$ occurred, because (*a*) we have decided to use only occurrence or non-occurrence of $R$, and (*b*) the fact that $R$ occurred supports $H$ much less well than some rival. (*a*) is a decision about what experiment to make, while (*b*) is straightforward likelihood testing.

So when suitable for after-trial evaluation, the Neyman–Pearson theory is part of the likelihood theory.

What about (a), the decision to garner only part of the available data? This too falls in the province of likelihood theory. For in deciding which data to collect we are essentially considering 'meta-hypotheses'. When statistical hypothesis $H$ is under test, and the initial information $d$ is given, we can compare how well $d$ supports the rival meta-hypotheses $h$ and $i$: of which $h$ asserts that limiting our observations to $R$ will lead us treat $H$ wrongly, whereas more extensive observation would get it right; the rival $i$ says that we are just as well off with $R$ as with more expensive experimenting.

Although these rivals $h$ and $i$ can be compared in terms of support, a practical man of business wants more. He is concerned with costs. He will lose a lot if he is wrong about $H$, but he does not want to spend too much discovering whether $H$ is true. The balance of these factors must be found within decision theory, and so is beyond the scope of the present inquiry. But I may take the opportunity to remark on the two conflicting emotions which move any serious student of current decision theory. On the one hand must be his respect for the elegant mathematical structures which have been developed from the fundamental insights of Wald and von Neumann. Yet he must often be assailed by sceptical doubts about a few seemingly facile assumptions which so often pass unnoticed: especially the assumption that all action should be guided by the mathematical, or actuarial *expectation*.

*Some other views*

Despite general acceptance, the Neyman–Pearson theory has, ever since its inception, been subject to severe criticism. Fisher has been the most ardent polemicist, and I think that the above conclusions do coincide with Fisher's intuitions. He was certain that the Neyman–Pearson theory is at most applicable to some kinds of routine and repeated investigation, and that it does not pertain to evaluating statistical hypotheses in the light of all the data available. He is most sceptical of Neyman's use of tests which do, from his point of view, simply discard data. Moreover the tests which he has favoured generally turn out to be likelihood tests. I hope that describing their structure within the theory of support may contribute to an understanding of their logic.

## SOME OTHER VIEWS

The idea that Neyman–Pearson tests could better serve before-trial betting than after-trial evaluation has lain fallow in a remark made by Jeffreys twenty years ago. Last year Dempster produced, for the first time in all that period, a similar distinction; much the same kind of idea is found in an article of Anscombe's published at about the same time.† In all these works it is suggested that however adequate be the Neyman–Pearson theory for what I have called before-trial betting, it is suitable for after-trial evaluation only in the light of a decision to discard or ignore some of the available data. The present chapter may serve to present these ideas within a single and uniform theory of statistical support.

### Neyman and Pearson's own rationale

It is of some interest to study the original rationale for Neyman–Pearson tests. In using a test, say Neyman and Pearson, there are two possible kinds of error. First, one might reject $H$ when it is true. Secondly, one might fail to reject $H$, when a rival $J$ is true. According to them, one should set the chance of an error of the first kind at some small figure, and then choose $R$ so as to minimize the chance of an error of the second kind. Evidently this means maximizing the power of the test for a given size.

Small size and large power are taken to be intrinsically desirable. Pearson called this 'the mathematical formulation of a practical requirement which statisticians of many schools of thought have deliberately advanced'.‡ But these two authors, especially Neyman, have a general rationale for the point of view. It is more radical than anything I have mentioned so far. They entirely reject the idea of testing individual hypotheses. As their 1933 paper to the Royal Society puts it,

> We are inclined to think that as far as a particular hypothesis is concerned no test based upon the theory of probability can by itself provide any valuable evidence of the truth or falsehood of that hypothesis.

† H. Jeffreys, *Theory of Probability* (Oxford, 1939), p. 327. A. P. Dempster, 'On the difficulties inherent in Fisher's fiducial argument', read at the International Statistical Institute in Ottawa, 1963. F. J. Anscombe, 'Sequential medical trials', *Journal of the American Statistical Association*, LVIII (1963), 365–83.

‡ E. S. Pearson, 'The choice of statistical tests illustrated on the interpretation of data classed in a $2 \times 2$ table', *Biometrika*, XXXIV (1947), 139–63.

Here follows a footnote,

Cases will, of course, arise where the verdict of a test is based on certainty. The question, 'Is there a black ball in this bag' may be answered with certainty if we find one in a sample from the bag.

Then in the body of the text,

Without hoping to know whether each separate hypothesis is true or false, we may search for rules to govern our behaviour with regard to them, in following which we will ensure that, in the long run of experience, we shall not be too often wrong. Here, for example, would be such a rule of behaviour: to decide whether a hypothesis $H$, of a given type, be rejected or not, calculate a specified character, $x$, of the observed facts; if $x > x_0$ reject $H$; if $x \leqslant x_0$, accept $H$. Such a rule tells us nothing as to whether in a particular case $H$ is true when $x \leqslant x_0$ or false when $x > x_0$. But it may often be proved that if we behave in such a way we shall reject when it is true not more, say, than once in a hundred times, and in addition we may have evidence that we shall reject $H$ sufficiently often when it is false.

I know of no better statement of their position than this, their first. It is very generally accepted. In the eyes of these authors, there is no alternative to certainty and ignorance. They say that 'no test based on the theory of probability can by itself provide valuable evidence of the truth or falsehood of an hypothesis', and, in a footnote, indicate that what they mean by valuable evidence is no less than certain proof.

These men propose that we should not hope to have 'valuable evidence' about any particular hypothesis, but that we should consider the whole class of hypotheses which we shall ever test. It is important to distinguish different possibilities. One might imagine us going on and on, testing the same hypothesis, resolutely forgetting what we had learned from one test as soon as we passed to the next—this makes a trifle more sense if one thinks of a host of independent experimenters working on the same chance set-up under conditions of total secrecy. It might be claimed that it can be 'proved', as Neyman and Pearson say, that if everyone uses the same test, not more than one in a hundred experimenters will mistakenly reject the true hypothesis. But of course this cannot be proved. It cannot be proved of any finite group of experimenters, or any finite sequence of trials. It can at most be proved that there is an extremely good chance that not more than one in a hundred experimenters will go wrong. So before any actual trials are made, there is exceptional support for the hypothesis that only 1 % will err. But this is not certainty. It has sometimes been called 'practical' or 'moral' certainty but that is just a tedious way of concealing the fact that one has given up certainty and resorted to the logic of support.

In their article our authors do not have in mind many different tests on the same hypothesis, but single tests on different ones. The situation is no better. It cannot be proved 'that if we behave in such a way we shall reject when it is true not more, say, than once in a hundred times'. As far as, in their words, the '*long run of experience*' goes, all that can be proved is that there is a supremely good chance that this will happen. Hence in defending their theory, Neyman and Pearson must admit that there is at least one hypothesis for which there can be evidence short of certainty, namely the hypothesis that 'in the long run of experience' a certain procedure will lead one to reject wrongly no more than one time in a hundred.

A few writers have already voiced a fear that the Neyman–Pearson theory defends chance in terms of chance. To stop that spiral, one needs the logic of support. The inductive behaviour which Neyman counsels cannot be proved to have the property he desires. But it can be proved that possession of this property is a very well-supported meta-hypothesis. It can be proved if one admits the law of likelihood. That is Neyman's dilemma: either admit something like the law of likelihood, or admit his theory unfounded. But admitting the law of likelihood means admitting evidence short of certainty. It also means admitting the possibility of tests like likelihood tests.

Fisher battered away at the Neyman–Pearson theory, insisting that even if it worked for testing long sequences of hypotheses, as in industrial quality control, it is irrelevant to testing hypotheses of the sort important to scientific advance. Even Pearson admitted qualms on this point, but did not surrender. But this is not my present criticism. Even when a long run of hypotheses are actually to be tested, something like the law of likelihood is required to validate the Neyman–Pearson reasoning.

But let us grant enough of the theory of support to the Neyman–Pearson theory. Suppose I had to adopt a testing policy today, and that I was bound to it for the rest of my life. At life's end I reap my rewards. For every false hypothesis I reject, I collect $r$ units of eternal utility, and $r'$ for every true one I fail to reject; for every true one I reject, I lose $s$, and lose $s'$ for every false one I fail to reject. I can choose whatever testing policy I like. Under these circumstances, the most glorified of all before-trial betting situations, I should opt for an optimum Neyman–Pearson test, and even favour unbiased tests at least to the extent that I wished to avoid 'worse than useless' tests. Very well. But no one has ever been in this situation. Perhaps it has seemed that this fact is irrelevant, on the grounds that the best policy for this situation is the best policy for after-trial evaluation. Preceding examples prove these 'grounds' are simply mistaken.

The Neyman–Pearson theory has had a long run of being regarded as a successful and complete theory of testing. I do not think many statisticians have taken very seriously Neyman's claim that one should adopt a life-long policy, an 'extended long run justification'. I am inclined to suppose that the theory's popularity has been in part due to the habit of statisticians of not bothering to state very explicitly the data upon which a rejection should be based: those who have insisted upon it, like Jeffreys and Fisher, have, from their very different standpoints, always been extremely sceptical and critical. We have shown that the theory is a perfectly adequate theory for after-trial evaluation, *if accompanied by a rational decision to discard or ignore data*. Most Neyman–Pearson theorists have inclined to forget that they do discard or ignore data in using Neyman–Pearson tests; it has been all the easier to do this because in some cases likelihood and Neyman–Pearson tests coincide, and so use of the Neyman–Pearson test does not always involve ignoring data.

*Sequential tests*

We have been speaking as if the only possible consequence of a test is that one might reject or fail to reject an hypothesis. This has good historical precedent, but it is pretty parochial today. In reality, if $H$ is being tested against some $J$, possible results of a test divide into three categories: reject $H$; accept $H$ and so reject $J$; and remain agnostic. The third possibility invites further experimentation. It suggests a theory of *sequential tests*; you make a trial, and reject, accept, or remain agnostic; in the third case you must

## SEQUENTIAL TESTS

make another trial, with the same three possible outcomes. Hence one is led to study the properties of a sequence of tests. The first investigation of the general theory is due to Wald;† his tests are essentially, or can be transformed into, sequences of likelihood tests. Far more constraints may be imposed on the problem than those I have described. There may, for instance, be a fixed cost of experimentation, so it is a problem to discover the chance, before the experimenter sets out, that his resources will be squandered before he either accepts or rejects the hypothesis under test. This is an interesting variant on the classical gambler's ruin problem, when it is a problem to discover which of two opposing gamblers is most likely to go bankrupt.

Incidentally, even though it is not necessarily desirable to maximize the power of a test at the first or second or third stage, on the other hand, it may seem desirable that if an hypothesis is false, the chance of its being rejected should increase with more and more experimentation. Under very general conditions likelihood tests do have this pleasing property: their power actually approaches 1 as more and more trials are made. They have many other pleasant asymptotic properties: they are, for instance, asymptotically unbiased.‡ But there is no reason, as my earlier example has shown, to expect unbiasedness at some early stage in the experiment.

### *Optional stopping*

The idea of sequential tests suggests a new difficulty. A man who follows a sequential policy has the option of stopping his experiment when he has enough data to settle the question which concerns him. Does this affect the logic of his tests? It may seem to. Let us put side by side two different cases.

Abel is testing simple hypothesis $h$ against a simple rival $i$. He decides in advance to make $n$ trials of kind $K$, or in other words to make one compound trial of kind $K_n$. He agrees that $h$ should be rejected if the likelihood ratio of $i$ to $h$ exceeds $\alpha$, more precisely, if he observes a result $E_n$ such that,

† A. Wald, *Sequential Analysis* (New York, 1947).
‡ A. Wald, 'Tests of statistical hypotheses concerning several parameters when the number of observations is large', *Transactions of the American Mathematical Society*, LIV (1943), 426–82.

$$L(E_n) = \frac{\text{Chance of getting } E_n \text{ on a trial of kind } K_n, \text{ if } i \text{ is true}}{\text{Chance of getting } E_n \text{ on a trial of kind } K_n, \text{ if } h \text{ is true}}$$

exceeds $\alpha$. He then makes a trial of kind $K_n$, observes a result $E_n$ such that $L(E_n) > \alpha$, and concludes by rejecting $h$.

But now consider Cain, who abhors $h$ and strongly favours $i$. He decides to continue making simple trials of kind $K$ until $i$ is sufficiently well supported that he can reject $h$ at critical ratio $\alpha$. If he never gets such support before his resources are squandered, he will simply not publish his work. In fact Cain makes exactly the same experiment as Abel, observes $E_n$, computes $L(E_n)$, finds it exceeds $\alpha$, rejects $h$ for all the world to see, and stops.

Cain may seem to be following a vicious sequential policy. For, it may be protested, he computed $L(E_n)$ as if he were just making a trial of kind $K_n$. But in fact he was making trials of quite a novel kind, sequential trials, which we shall call trials of kind $M$. First, let us call 'success' the event that the man makes $m$ trials (for some $m \geq 1$) and observes a result $E_m$ such that $L(E_m)$ exceeds $\alpha$. We shall say that an $m$-success occurs if success occurs at the $m$'th trial of kind $K$, but has occurred at no earlier trial. Our man is making trials of kind $K$ until he observes success. Thus he can be said to be making trials of kind $M$, trials of which the only possible results are $m$-successes, namely 1-success, 2-success, and so on. Cain observed $n$-success. In computing the relative support for $h$ and $i$ he must not compute $L(E_n)$, but

$L(n\text{-success}) =$

$$\frac{\text{Chance of getting } n\text{-success on trials of kind } M, \text{ if } i \text{ is true}}{\text{Chance of getting } n\text{-success on trials of kind } M, \text{ if } h \text{ is true}}.$$

Might it not happen that $L(n\text{-success})$ differs from $L(E_n)$?

Two separate intuitions bear on this objection. First, it can be proved that regardless of which of $h$ or $i$ is true, and no matter how high $\alpha$ be set, Cain must score a success sooner or later; more correctly, that the chance of his doing so is exactly 1. Hence our vicious experimenter who hates $h$ is bound to reject it sooner or later by following the policy I have described. Thus, one is inclined to intuit, his actual experimental results are virtually worthless, or at least certainly must be reconstrued.

But there is a quite opposite intuition. Suppose Abel and Cain are colleagues. Together they make exactly $n$ trials of kind $K$ on their equipment. But Abel had privately decided from the beginning to stop after $n$ trials; he did not tell his colleague for fear of arousing antagonism. Cain had decided to follow the vicious policy, but has kept it secret. They both observe exactly the same sequence of events. Could it be possible that, on this data, the one experimenter who had settled $n$ in advance is entitled to reject $H$, while the other experimenter is not? Can testing depend on hidden intentions? Surely not; hence optional stopping should not matter after all.

The solution to the difficulty lies in the facts. Although the chance of getting $n$-success, if $h$ is true, differs from the chance of getting $E_n$ on trials of kind $K_n$, if $h$ is true, the likelihood *ratios*, namely $L(E_n)$ and $L(n$-success), turn out to be identical. *Likelihood ratios are insensitive to optional stopping.* Hence although it may be morally deplorable to pretend one had settled $n$ in advance, when one had not, such a lie is statistically innocuous. Hence the second intuition just stated turns out to be right. What of the first? Well, it is true that the chance of scoring success *sometime* on an unending sequence of trials will be 1, whether or not $h$ is true. But no one makes unending sequences of trials. If a man has scored success in the first $n$ trials, what is of interest is his chance of doing exactly that, first on hypothesis $h$, and then on hypothesis $i$. And this gives us the same likelihood ratio as $L(E_n)$.

*Support and likelihood tests*

Now for some general advantages of likelihood tests. They apply to all distributions for which density functions exist, and hence to all families of distributions whose empirical significance can be readily understood at present. Nor is there any need for an unending search for conditions to guarantee that likelihood tests are unique. The really standard tests agreed to by statisticians of all schools turn out to be likelihood tests. Unlike Neyman–Pearson tests, they are insensitive to optional stopping.

The theory of likelihood tests stands entirely on the logic of support. One might venture an abstract principle about testing in general: a good test for any hypothesis, statistical or otherwise, is one which rejects it only if another hypothesis is much better supported. This principle, plus the logic of support and the law of likelihood entails the theory of likelihood tests.

A theory of rejection based on the logic of support ties in with a thesis very commonly accepted. Most statisticians hold that a test based on sufficient statistics is as good as a test based on all statistical data. Sufficient statistics are defined as follows: for a class of hypotheses $H$, a statistic $d$ is a sufficient statistic of, or contraction of, statistical data $e$ if and only if for every pair of hypotheses $h$ and $i$ in $H$,

$$\frac{\text{Likelihood of } h, \text{ given } d}{\text{Likelihood of } i, \text{ given } d} = \frac{\text{Likelihood of } h, \text{ given } e}{\text{Likelihood of } i, \text{ given } e}.$$

Now why should sufficient statistics, as so defined, be as good a basis for a test as more complete statistical data? That they are as good has even been canonized as a *principle of sufficiency*.† I. J. Good suggests that the only reason to be given for this principle is that Fisher so often implied it.‡ Some workers take it as an accidental consequence of a theory of testing, while others defend it in terms of an analogy with gambling.§

One ground for the principle of sufficiency is evident within the logic of support. Hypotheses are to be rejected only if a rival is better supported. Support is analysed in terms of likelihood ratios. Sufficient statistics are those which leave likelihood ratios constant within a class of hypotheses. Hence sufficient statistics are as good a basis of tests, within the given class of hypotheses, as the complete statistical data.

*Objections to likelihood tests*

At least three separate objections might be raised. First, following Neyman's opinions quoted above, you might argue that there can never be statistical evidence against a unique hypothesis taken alone; at most the statistician can adopt a policy for testing a whole sequence of hypotheses. It is because of this objection that I was at pains to prove how even Neyman's theory must sooner or later invoke something like the law of likelihood. This is Neyman's dilemma: either abandon his own theory, or else admit evidence short of certainty for at least one unique statistical hypothesis.

† A. Birnbaum, 'On the foundations of statistical inference', *Journal of the American Statistical Association*, LVII (1962), 270.
‡ I. J. Good, *ibid*. 'Discussion', p. 312.
§ P. R. Halmos and L. J. Savage, 'Application of the Radon–Nikodym theorem', *Annals of Mathematical Statistics*, XX (1949), 240.

Secondly, it may be said that likelihood tests have no clearly defined optimum properties. They need not be, for instance, UMP unbiased or have any of the other properties which have been thought optimum. To this must be retorted that it is a question of what is optimum. When no test is uniformly most powerful, it is not evident that being more powerful than some other test makes a test better; my preceding counter-examples prove this need not be so. It is not clear what properties likelihood tests 'ought' to have. They do have the splendid property of rejecting only those hypotheses which are least well supported.

Thirdly, there may be counter-examples to the excellence of likelihood tests construed as tests of unique hypotheses. This is the best way to refute a principle: not general metaphysics but concrete example. One family of counter-examples is known: the 'worse than useless' tests. My own counter-example to Neyman–Pearson invariant tests is an extreme instance. All rely on finding a likelihood test which is demonstrably not the most powerful. But in none of the so-called counter-examples is it demonstrable that this is a defect in a test for after-trial evaluation. An earlier example shows that at least one 'worse than useless' test is perfectly sensible.

*The critical ratio*

Roughly speaking a likelihood test with critical ratio $\alpha$ rejects an hypothesis $h$ on data $e$ if there is an hypothesis $i$ consistent with $e$, and such that the likelihood ratio of $i$ to $h$ exceeds $\alpha$. So the critical ratio of a test is a measure of its stringency. Any hypothesis which would be rejected on some data by a test of given critical ratio will be rejected on the same data by a yet more critical test. The question remains, how stringent should a test be? There is no more a general answer to this question than there is to the question whether to reject a scientific theory which has powerful but not absolutely compelling evidence against it.

What one can do is to compare the stringency of different rejections. Now if statistical data were the only possible data bearing on hypotheses, and the only grounds governing choosing between them, it might be reasonable to insist on a unique level of stringency. For one might argue, if you reject this hypothesis at this level of stringency, you must reject that one anyway: it would somehow seem unreasonable to have different levels of stringency. I cannot see how to prove it would be unreasonable. But one might simply assume it as unreasonable, and take the level of stringency as a characteristic of individual persons. A reasonable person who 'jumped to conclusions' would be one with a low critical ratio in likelihood tests; a pedantic phiilosopher ever eyeing remote possibilities would be one with an exceptionally high critical ratio. But all this is in a vacuum. Statistical data are never the only data bearing on a practical problem. 'All things being equal except the statistical data' is empty.

It follows that choice of the critical ratio depends on matter extraneous to the statistical data. This matter is of two kinds. First, there may be prior but not statistical data bearing on several of the possible hypotheses, and already inclining the mind for or against some or other of these. Secondly, some hypotheses may have consequences far more grave than others, and so of themselves demand a greater stringency.

There is an extreme case of prior data: when it is known that the hypothesis under test is itself a possible outcome of a chance set-up whose distribution is known. This sort of problem was first considered by Bayes, and his reflexions were published by the Royal Society in 1763. I shall show in the sequel how Bayes' problem may be reduced to a problem in the logic of support.

Aside from this extreme, there is no obvious way numerically to evaluate prior data. However, there are a number of theories on how to use information about the relative gravity of different hypotheses. One author has suggested that a number be assigned to each hypothesis, which will represent the 'seriousness' of rejecting it.† The more prior information supporting the hypothesis, and the greater the gravity of rejecting it, the greater the seriousness of the hypothesis. In the theory of likelihood testing, one would use weighted tests. Thus, if simple $h$ were being tested against simple $i$ with critical ratio $\alpha$, and if the seriousness of $h$ were represented by the number $a$, and that of $i$ by $b$, $h$ would be rejected by data $e$ only if the likelihood ratio of $i$ to $h$ exceeded $a\alpha/b$.

A standard sort of example shows the merit of this proposal. Suppose $h$ entailed that a polio vaccine, administered to children,

† D. V. Lindley, 'Statistical inference', *Journal of the Royal Statistical Society*, B, xv (1953), 30–65.

would cause the same children to die of heart disease before they were twenty; and that the rival $i$ entailed that the vaccine did not have this evil consequence. If the vaccine were to be administered or not according to discoveries about $h$ and $i$, $h$ would be serious and $i$ less so. $h$ would be rejected for $i$ only if the likelihood ratio of $i$ to $h$ exceeded $a\alpha/b$, and since $a$ would be very large, $h$ would be rejected only if there were a lot of evidence against it. This seems desirable. But talk of seriousness at present can only be used to represent a kind of reasoning, and not, I should guess, to test any piece of inference.

There is, then, no clear-cut answer to the question of stringency. What stringency to choose, and how to weight accordingly, remain very largely a matter of individual judgement. One may compare judgements, but seldom can one settle definitively which judgement is correct. If one man applies a more stringent standard than another, rejecting what the other accepts, there is a well-known way to settle the dispute. Get more data.

Finally, let it never be forgotten that there are statistical hypotheses in a sufficiently limited domain which would never be rejected. As Gossett said, 'I can conceive of circumstances, such for example as dealing a hand of 13 trumps after careful shuffling *by myself*, in which almost any degree of improbability would fail to shake my belief in the hypothesis that the sample was in fact a reasonably random one from a given population'.†

*Refutation*

Rejection is not refutation. Plenty of rejections must be only tentative. Let it be supposed that the distribution in connexion with a coin tossing experiment is binomial—independent trials with the chance of heads between 0 and 1. The hypothesis under test is that $P$ (heads) exceeds $1/2$. Let a likelihood test be used, with critical ratio 100. Suppose the coin is tossed 9 times, and falls heads only once. A test with ratio 100 rejects the hypothesis under test. (The size of this test is 0·02.) But suppose on the very next toss the coin were to fall heads again. Our hypothesis should no longer be rejected; it is up for consideration again. Further data may reinstate what was once rejected. This is perfectly natural.

† Quoted in E. S. Pearson's memoir of Gossett (*Biometrika*, xxx, 1938, p. 243).

Braithwaite usefully compares rejection tests to procedures for putting hypotheses into two baskets, 'rejected' and 'up for consideration'. New data may move something from the first to the second.

Though some tests are tentative, not every one need be. It may seem as if any statistical hypothesis, rejected on any data, might be reinstated later. I do not think so. A coin falling 4999 out of 5000 times is pretty conclusive evidence that, under the conditions of tossing, it is not fair. Even if many heads occur in subsequent trials the hypothesis that the coin is fair under those conditions will not be reinstated. Rather, one will conclude that the experimental conditions have changed, and with them, the distribution.

Certainly 4999 tails out of 5000 consecutive tosses is formally consistent with the chance of heads being 1/2. But it is in some way incompatible with it. No future evidence, together with 4999 of the first 5000 tosses being tails, will ever support that hypothesis better than some rival hypothesis, say, that the distribution has changed, or, in Gossett's words, whatever will do the trick. It will never support $P(\text{heads}) = 1/2$, one-thousandth as well. This is why statistical hypotheses may be refuted. Indeed if one supposed that an hypothesis can only be refuted by observations formally inconsistent with it, then no statistical hypothesis can be refuted. But I take it that something is refuted if no matter what happens in the future, it will never be nearly as well supported as some rival hypothesis.

### Braithwaite's theory of chance

It is fitting here to consider the theory of Braithwaite's *Scientific Explanation* not because it is a contribution to the theory of testing but because it offers a postulational definition of chance in which the key postulates state when to reject a statistical hypothesis.

According to this theory, the meaning of statistical hypotheses is to be given by rules for rejecting them. There is a family of such rules. For any real number, $k$, a $k$-rule of rejection states, in our terminology, that on the data that in $n$ trials of kind $K$ on set-up $X$, the proportion of trials in which $E$ occurred exceeds

$$p+\sqrt{(p(1-p)/nk)} \quad \text{or is less than} \quad p-\sqrt{(p(1-p)/nk)},$$

then the hypothesis that $P(E) = p$ is provisionally to be rejected.

Evidently the larger the value of $k$, the stricter the corresponding $k$ rule of rejection. It is also obvious that on data about $n$ trials, a $k$-rule might instruct provisional rejection, while after $m$ trials more, the rule, considering data based on $n+m$ trials, might not instruct rejection. In that event, the statistical hypothesis, provisionally rejected after $n$ trials, would be reinstated after $n+m$.

This theory asserts that $k$-rules of rejection give the meaning of statistical hypotheses. But $k$-rules differ among themselves, according to the value of $k$. Hence different $k$-rules must attach different meanings to statistical hypotheses. Better, each $k$-rule defines what I shall call $k$-statistical hypotheses; instead of using the term 'chance' we may speak of $k$-chances, namely the chances mentioned in $k$-statistical hypotheses, and whose meaning is given by $k$-rules of rejection.

There is nothing intrinsically objectionable to a whole continuum of $k$-chances, where $k$ may be any real number. After all, we are to explicate our idea of a physical property, long run frequency; perhaps a continuum of possible explicata is necessary. This, at any rate, is Braithwaite's view of the matter.

Braithwaite's theory is, however, very odd. $k$-chances are supposed to be physical properties: $k$-statistical hypotheses ascribe physical properties to the world. For any $k_1 \neq k_2$, two possibilities emerge: (1) the $k_1$-chance of something must be the same as its $k_2$-chance, or (2) $k_1$ and $k_2$-chances may differ.

First take the former alternative. Suppose $k_1 > k_2$, so that the $k_1$-rule is the more stringent. Suppose that on some data, the $k_1$-rule instructs provisional rejection of the hypothesis that the $k_1$-chance of $E$ is $p$, while the $k_2$-rule does not instruct rejection of the corresponding $k_2$-hypothesis. All the same, it must be correct to reject the hypothesis that the $k_2$-chance is $p$, for on alternative (1), $k_1$ and $k_2$-chances coincide, so if you provisionally reject the $k_1$-hypothesis you should, in consistency, reject the $k_2$-hypothesis also.

Following this line of thought it is possible provisionally to reject virtually any statistical hypothesis. Consider the $k_2$-hypothesis that the $k_2$-chance of $E$ is $p$. Let there be any data, say that in $n$ trials there were $m$ occurrences of $E$. Then so long as $p \neq m/n$ you should reject the $k_2$-hypothesis, regardless of the value of $m/n$, or $p$, and of $k_2$. Simply take a large enough value of $k_1$, 'provisionally' reject the corresponding $k_1$-hypothesis about the $k_1$-chance of $E$ being $p$, and hence reject the $k_2$-hypothesis. Thus on any data, and any value $k$, you may provisionally reject every $k$-hypothesis but one. This vitiates Braithwaite's theory.

It may be thought that the 'provisionally' saves the theory. But it does not. For any theory which says that you must, on any data, provisionally reject all but one hypothesis, is simply vicious, and utterly distorts statistical inference.

Let us turn, then, to alternative (2): $k_1$ and $k_2$-chances of the same thing may differ. This alternative is just mysterious. What on earth must be the case for the $k_1$-chance of $E$ on trials of kind $K$ to be $p$ while the $k_2$-chance is not (really not, not just in our opinion not) equal to $p$? No remotely plausible answer comes to mind. Not if $k$-chances are physical properties, as Braithwaite contends.

It may be protested that we have raised a false dilemma. There are further options to (1) and (2). Perhaps it is not significant to ask whether $k_1$ and $k_2$-chances could be the same or differ. But if so, then the definition of $k$-chances is deplorable, for it fails to permit answering any important questions about identity.

The dilemma, bad (1) and worse (2), is a forceful way to bring out one error in defining chance in terms of rules for rejection. According to Braithwaite, we choose particular $k$-rules for particular purposes. If it is important not to fail to reject a false hypothesis, we use a strict $k$-rule with a large $k$; if not so important, while it is essential not to reject a true hypothesis, we choose a lower value of $k$. This is an admirable account if we are discussing the stringency of tests. But it is not so good if, as in Braithwaite's theory, the $k$-rules are supposed to define a property and to give the veritable meaning of statistical hypotheses.

Of course people are more or less stringent according to their immediate ends. But Braithwaite wrongly takes this to be a matter of meaning. He says 'the arbitrariness, from the logical point of view, of the choice of the number $k$ is an essential feature of the way we attach meanings to statements asserting probabilities [chances] which are neither 0 nor 1'. And we should take a low $k$ if the use of the hypothesis is to be 'purely intellectual'; a high one if the hypothesis would 'be of the greatest practical use to us' if true. But this is extraordinary. Compare (1) 0·6 is the long run frequency with which the waiting time for a radioactive emission from this material exceeds 1 sec., and (2) the long run frequency with which inoculated patients catch typhoid is 0·6. Or replace 'long run frequency' by 'chance' in those assertions. Suppose (1) is purely intellectual, and that (2) is of the greatest practical importance. Should we attach a meaning to chance, or long run frequency, in the first case different from that we attach in the second? I think not. Indeed we may be reluctant to reject the second hypothesis and eager to reject the first. More stringent tests may be applied to one than the other. But is this a difference in meaning? Suppose a poisonous mushroom has purple spots on the stem, as does an academically interesting poppy. Then, on a walk through the woods, the claim of a particular poppy plant that it has no purple spots on the stem may be 'purely intellectual', while the corresponding claim of a mushroom may be 'of the greatest practical importance'. At the least sign of purple spots on the mushroom I shall discard both plant and hypothesis, while when picking poppies I shall not be nearly so stringent. But a difference in the meaning of the phrase 'purple spots'? That would be absurd. It is no less absurd to imply that our private hopes and fears determine the meaning of 'chance' or 'long run frequency'.

CHAPTER VIII

# RANDOM SAMPLING

Many persons take inference from sample to population as the very type of all reasoning in statistics. It did lead to the naming of our discipline. 'Statistics' once meant that part of political science concerned with collecting and analysing facts about the state. 'Statistical inference' meant the mode of inference peculiar to that part of the science. The meaning has since been extended, but it is no verbal accident that 'population' is the name now given in statistics to a collection of distinct things which may be sampled.

What is the pattern of inference from sample to population? 'Make a random sample of a population, assume in certain respects that it is typical of the whole, and infer that the proportion of $E$'s in the whole population is about equal to the proportion in the sample.' That would be the most naïve account. It is grossly inadequate, especially if the idea of a random sample were to remain undefined. In fact the process of inference from random sample to population is entirely rigorous. We must first attend neither to sample nor population, but to the sampling procedure. Any such procedure involves a chance set-up. Once the analysis of random sampling is complete, inference from sample to population follows as a trivial corollary of the theory of chance. In addition, most of the hoary riddles about randomness can be solved or else shown irrelevant to statistics.

*Randomness*

Three questions about randomness are to be distinguished. (1) What does the English word 'random' mean? Perhaps this can be answered briefly, but it would take 100 pages to prove any answer correct. I shall not try here. (2) What is the appropriate mathematical concept pertaining to infinite series, which is most similar to the ordinary conception of randomness? This problem has been definitively solved by Church, but we shall not require his results.† (3) Which features of random samples are crucial to statistical inference? This is our question. We shall answer it in a sequence of distinct steps.

*Random trials*

It makes sense to describe trials on a set-up as random but in what follows I shall not do so because it is unnecessary. I take it that trials are called random if and only if they are independent. Hence it suffices to speak of independence.

The hypothesis of independence can be tested like any other in statistics. A few words on this score are worthwhile, for some well-known tests are often presented in a way which conceals their logic.

In general, if in some experiment there is a suggestion that trials may not be independent there will be some knowledge of what kind of dependence is to be feared, or the results themselves may suggest to the experimenter that a particular kind of dependence is at work. One well-known form of dependence involves frequent occurrence of 'runs', where by a run is meant a sequence of trials with exactly the same result. The hypothesis that there is this kind of dependence can be tested against the hypothesis of independence.

Practical manuals often include instructions for applying such a test, but they seldom mention hypotheses rivalling that of independence. So they make it look as if there were tests which do not require formulation of rival hypotheses. The appearance is deceptive. The test is good only against some class of rival hypotheses. This class includes many forms of dependence found in practice, so it is worth testing one's laboratory trials to see if they display one of these forms. Another well-known test involves computing the proportion of times in which the observed results have a value above their average value. This test is effective in testing against a somewhat different class of hypotheses. So it may be wise to apply this test as well as the preceding one. An experimenter might apply both tests blindly because he is told they are good, and never formulate any rival hypotheses, but his tests are good because they are good against certain kinds of rivals.

† A. Church, 'On the concept of a random sequence', *Bulletin of the American Mathematical Society*, XLVI (1940), 130–5.

*Random sequences*

Perhaps it makes sense to call a sequence random but I shall not do so. The expression is nearly always misleading. Church has been able to give precise sense to the notion of an infinite random sequence, but his analysis does not apply to the finite sequences which will occupy us here.

Random sequences are often conceived to be sequences which display no discernible order. Since one can discern some sort of order in any finite sequence, the only surprising thing about the failure of this idea is that some authors have hoped to salvage a precise notion from it.

According to von Wright, 'the distribution of $A$ in $H$ is called random if there is no property $G$ such that the relative frequency of $A$ in $H \& G$ is different from the relative frequency of $A$ in $H$'; where he has explained that by 'no property' he means no property except properties entailing the presence of $A$.† But take any property $B$ which as a matter of contingent fact is never found with $A$ (or never with $A \& H$). Then being $A$-or-$B$ does not entail, in von Wright's sense, being $A$, yet the relative frequency of $A$ in $H \& (A$-or-$B)$ is necessarily 1. So on von Wright's account the distribution of $A$ in $H$ would be random only if all $H$'s were $A$.

If one were to insist on speaking about random sequences, I think one should intend no more than sequences of things taken at random. In particular, a sequence of results from trials on a chance set-up would be called random if and only if the trials are independent. Random sequences would be the sequences provided by random trials. This does not entail there is 'no discernible order' but only that if the sequence displays an order which establishes to everyone's satisfaction that the trials were not independent, and if there are no other data about the set-up, then the sequence will not be called random.

Such an account of randomness does not deny that the idea of a sequence with 'no discernible order' is a pleasant source of puzzles. You can try to invent systems for producing unending sequences of digits in which no one, not informed of the system, could discern any order. Popper has some interesting examples in

† G. H. von Wright, *A Treatise on Induction and Probability* (London, 1951), p. 229.

this genre.† Here is an amusing problem suggested by Cargile: produce an unending sequence of digits no part of which ever repeats, that is, within which there is no sequence of $n$ consecutive digits immediately followed by exactly the same $n$ digits. If non-randomness were construed in terms of direct repetition, such a sequence would certainly be random. To make such a sequence you need four numerals. Note that the following four sequences are non-repeaters:

$s1$:  1 2 3 4
$s2$:  1 3 2 4
$s3$:  1 4 3 2
$s4$:  1 4 2 3

Let $ss1$ be the 16 termed sequence formed by replacing 1 in $s1$ by $s1$ itself, 2 by $s2$, 3 by $s3$, and 4 by $s4$. Then $ss1$ is a non-repeater. So is $ss2$, which is constructed by substitution in $s2$. $sss1$ will be the result of replacing 1 in $s1$ by $ss1$, etc. $ssss...sss1$, constructed in like fashion, is a non-repeater. And we have a system for constructing a sequence which contains no direct repetition. Write down 1, continue to $s1$, continue to $ss1$, and so on. All the same, every fourth digit is a 1. Popper's sequences avoid this feature, but they are not non-repeaters in the above sense. These larks do not seem to have very much to do with the randomness which concerns statistical inference.

## Two kinds of sampling

We can now turn to sampling. By sampling a population of definite and distinct things I shall mean singling out a member of the population, or else a sequence of such singlings out. This definition is used because it covers two kinds of sampling. They are most easily explained by example. Drawing balls from a bag is a way of sampling the contents of the bag, and there are two kinds of drawing. In *sampling with replacement* a ball is drawn, and is replaced before any more draws are made. In *sampling without replacement* drawn balls are not replaced. Thus in sampling with replacement the same individual may be singled out more than once while in sampling without replacement this cannot happen.

† K. Popper, *Logic of Scientific Discovery* (London, 1959), §54 ff.

## Two kinds of population

Some populations may have a definite and more or less practically ascertainable number of members; they will be called *closed*. In contrast are *open* populations whose members have in some sense not been determined, or, more correctly, whose members could in no way be determined at present nor have been determined in the past. The current membership of a society for vegetarians is a closed population, but the class of past, present and future English vegetarians is open. Thus explained, the notion of being closed is far from precise, but we shall consider only populations which are plainly closed.

Since a population is a class of distinct things, infinite populations, such as the population of natural numbers, do make sense. But they will not be considered in the sequel. I am not sure that it makes sense to speak of sampling a population of imaginary things, unless what is meant is something like writing down the names of Cerberus, Bucephalus and Pegasus or else drawing the names from a hat. At any rate we shall not consider sampling infinite populations most of whose members are imaginary; in the same spirit we shall pass by the hypothetical infinite populations which Fisher believed to be the foundation of all statistics.

## Random sampling

Suppose there is a chance set-up for sampling with replacement a closed population. Each trial on the set-up consists in singling out a member of the population. Suppose trials are independent. If the chance of singling out any one individual on any trial equals the chance of singling out any other, such a device will be called a *chance set-up for random sampling with replacement*. More briefly, call it a set-up for random sampling, or a random sampling device. A sequence of results from consecutive trials on such a device will be called a *random sample* of the population in question.

This definition is far from novel. It was given long ago by Peirce:

I mean by a random sample, a sample drawn from the whole class by a method which if it were applied over and over again would in the long run draw any one possible collection of the members of the whole collection sampled as often as any other collection of the same size, and would produce them in any order as often as any other.†

It is not altogether clear whether Peirce had in mind random sampling with or without replacement; probably the latter, in which case his definition is identical with the one given in the following section, rather than with what is defined above.

It will be observed that random samples are defined entirely in terms of the sampling device. Hence the hypothesis that a sample is random may be tested statistically. For example, if the sample furnishes evidence that the trials were not independent, it furnishes evidence that it itself is not random. But statistical methods are not the only way of proving non-randomness. If there are $n$ members in the closed population, the chance of singling out one member at any trial should be $1/n$. On occasion it might be shown without the aid of statistics that this condition is not met. To take a notorious historical example, suppose someone sampled the population of New York City by sticking pins in a telephone book, and then selecting the persons whose names he stabbed. The chance of picking out someone not in the directory is zero. Hence the persons he chooses do not form a random sample of the city's inhabitants. This does not mean the sample is useless in inferring properties about New Yorkers, but only that the sample is not random in the sense defined.

## Random sampling without replacement

A chance set-up which singles out a member of a population, then another, then another, and so on until trials cease or the population is exhausted, will be called a set-up for sampling without replacement. The device is random if at each trial the chance of singling out any member not hitherto selected is equal to the chance for any other such member. A sequence of results of trials on such a set-up will be called a *random sample without replacement*.

A sample consisting of the different individuals $E_1, ..., E_s$ might come from sampling with or without replacement; the sample $E_1, ..., E_1, ..., E_s$ could come only from sampling with replacement. However, if the population is large and the sample small, the chance of making any given sample in sampling without replacement is about the same as in sampling with replacement. If a fairly

† C. S. Peirce, *Collected Papers*, VII, p. 126; see also II, p. 454.

large proportion of the whole population has property $E$, the chance that a given sample has $k$ members with property $E$ will be about the same, for both types of sampling. So the chance properties of one kind of sampling approximate, in certain respects, the chance properties of the other.

In practice the most familiar kinds of devices sample without replacement. But the principles of sampling with replacement are much more readily explained than those of sampling without replacement. So in this chapter we shall consider only sampling with replacement, and treat sampling without it as an approximation to this. We have no theoretical reason; it is solely a matter of convenience.

*Inference from random sample to population*

Suppose that in a closed population of $n$ members, $k$ members have property $E$. Then on a single trial on a chance set-up for random sampling of this population with replacement the chance of getting an individual with property $E$ is exactly $k/n$. This is a consequence of our definition.

*This elementary fact conveys the principle of all inference from random sample to closed population.* We have already a general theory for testing the hypothesis that $P(E) = p$; this is automatically a theory for testing whether $k = np$. Likewise subsequent chapters provide theories for estimating the true value of $P(E)$; they are thus theories for estimating the proportion of $E$'s in the population. A host of practical problems remains; a few will be sketched forthwith. But the fundamental principle is already stated.

*How a sample can be typical of a population*

The naïve theory of inference from sample to population holds that the sample is in some way typical of the population, and hence that if the proportion of $E$'s in the sample is $p$, this is the proportion in the population as a whole. Unfortunately this form of words can describe two distinct kinds of inference only one of which is statistical in nature.

On the data that in a random sample from a population the proportion of $E$'s is $p$, the best supported hypothesis about $P(E)$ is $P(E) = p$, and hence the best supported hypothesis about the proportion of $E$'s in the population is $p$. In fact many theories say $p$ is the best estimate of $P(E)$, and hence the best estimate of the proportion. This indicates a sense in which it may be right to take the sample as typical of the population; the proportion in the random sample indicates the proportion in the population at large. But this is not a matter of assuming anything; it is a trivial consequence of the logic of statistical support. It is strongly to be contrasted with another course of inference.

Imagine as before that the population of a city is sampled by making a random sample from the names in the telephone book. This is not a random sample of the population at large. Suppose the proportion of persons with property $E$ in the sample is $p$, so that we estimate the proportion among persons listed in the directory as $p$. Now there may be reason to suppose that the proportion of persons listed, who have property $E$, is about the same as the proportion in the entire city. So we infer that $p$ is a good estimate for the city as a whole. Here too it may be said that we end by assuming the sample is typical of the population. But evidently a new kind of premise has been introduced, the premise that one population is typical of another. So we cannot regard this as direct inference from random sample to population sampled.

In the case of a city, economy might force you to sample the directory, but there is no reason in principle why you should not sample the population as a whole. But if you are concerned with the open population of seeds produced by some species of plant, and you have many thousand seeds of this sort in your experimental station, you might make a random sample of the seeds in your station, infer something about the population of seeds you possess, and then suppose your station's seeds are typical of the open population. If this is the inference you desire to make, you are forced, as a matter of logic, to make some extra-statistical assumptions. For there is no such thing as a random sample from the open population. The chance of sampling one of next year's seeds this year is necessarily zero. You must assume your laboratory's seeds are in some way representative of the open class of seeds of this sort; this assumption is very much like inferring a property of common salt from the fact that salt in your laboratory behaves in a certain way. It is commonly called induction, so the inference shall be called inductive inference. There is no known satisfactory analysis of inductive inference. But it is worth warning that both statistical and inductive inference could be naïvely expressed by the words, 'we assume the sample is typical of the whole'. Perhaps part of the inclination to try to justify all inductive inference by statistical methods stems from failing to see that two different things may be referred to by those words.

*Misleading random samples*

On the data that the proportion of $E$'s in a random sample from a population is $p$, the best supported hypothesis about the proportion of $E$'s in the whole population is also $p$. But evidently this hypothesis might not be supported by more inclusive data. In particular, there may be good reason to suppose the sample misleading. We shall examine the simplest instance of this sort of thing.

Suppose there is a random sample of the entire population of New York City in which it is observed that, by what might be called a fluke, all the individuals actually singled out are listed in the telephone directory. Suppose further that there is excellent reason to believe that the proportion of $E$'s among those listed in the directory is substantially greater than that in the population at large; $E$ might be the property of voting Republican. Then if $p$ were the observed proportion in the sample, everyone would doubt that $p$ is the proportion in the whole population. It is useful to see how the theory of support analyses this doubt. Except in cases of exceptional data, the theory cannot say what hypothesis is best supported by the whole of the data just described, but it can at least make plain that $p$ is not best supported. I do not believe any other theory of statistics can do better; perhaps on such data it is just not possible to determine which of several hypotheses is best supported.

Let $B$ represent the event that a person singled out should be listed in the directory. We shall begin by evaluating a rather precise piece of information, that the proportion of $E$'s among the $B$'s exceeds by at least $r$ the proportion of $E$'s among the whole of the population. That implies that in random sampling,

$$P(E|B) \geq P(E)+r,$$

where $P(E|B)$ is, as usual, the chance that an individual singled out is an $E$, given that it is a $B$. Thus our data reduce to two assertions: (i) Every member of a certain random sample from the population is a $B$, and the proportion of $E$'s in the sample is $p$; and (ii) $P(E|B) \geqslant P(E)+r$.

If the sample were large there would be overwhelming evidence that the data are incorrect: either this is not a random sample, or the inequality (ii) does not hold after all. Then the sample would not be called misleading; rather, the data would be suspected of falsehood. But let us examine the more instructive case, in which the sample is small and the data are regarded as unimpeachable.

How are we to use both (i) and (ii) in the theory of support? We can only do so by considering four possible results of a single trial: the event that the individual singled out is both an $E$ and a $B$, is an $E$ but not a $B$, $B$ but not an $E$, and finally, neither $E$ nor $B$. We consider the possible distributions of chances among these four outcomes, setting $P(EB) = a$, $P(E\bar{B}) = b$, $P(\bar{E}B) = c$, and $P(\bar{E}\bar{B}) = 1-a-b-c$. Notice that, in virtue of Kolmogoroff's axioms, $P(E) = a+b$, and $P(B) = a+c$. Moreover, in virtue of the definition of conditional chance

$$P(E|B) = \frac{P(EB)}{P(B)} = \frac{a}{a+c}.$$

Fact (ii) excludes some possible values of $a$, $b$ and $c$. According to it, these three fractions must satisfy the inequality

$$\frac{a}{a+c} \geqslant a+b+r.$$

Thus (i) and (ii) are equivalent to statistical data asserting that the possible distributions satisfy this inequality, and that in a sample, all the members were $B$ and a proportion $p$ were $E$.

The hypothesis about $a$, $b$ and $c$ best supported by (i) and (ii) will be that which maximizes likelihoods, and satisfies the above inequality. The solution is hideous to write down, and is of no matter. Suffice to say that on data (i) and (ii) taken together, the best supported hypothesis about $P(E)$ (namely about $a+b$) is not $p$, but something smaller.

In practice there would probably be data less explicit than (ii), and also more data like (ii). However, the point is already made: available data may indicate that the sample is misleading and is not typical of the population; they may indicate, for instance, that if the proportion of $E$'s in the sample is $p$, it is not true that the best supported hypothesis about the proportion in the population is $p$. In such a situation I call the sample *misleading*. The sample is misleading not in the sense that the proportion in the population simply differs from that in the sample, but in the sense that the available data do not give $p$ as the best supported hypothesis about the proportion in the population. As I use the term, to call a sample misleading is not to question its randomness. A sample fails to be random if it is not selected by independent trials. But a sample selected by independent trials can still be misleading. So there are two distinct ways of criticizing an inference from sample to population: (1) the sample is not random; (2) available data show the sample is misleading. You can go wrong either through dependent trials, or by the bad luck of drawing a misleading sample.

To show a sample misleading you need more than the premise that all the members of the sample have property $B$, and that only a part of the whole population has property $B$. For of course in any sample short of the whole population we can find a property possessed by all members of the sample, and not by every member of the population. If the sample consists of the individuals $A_1, ..., A_s$, then the property is 'being $A_1$ or ... or $A_s$'. Call this property $B$ if you like, but nothing follows; there is so far no reason to suppose that $P(E|B)$ differs substantially from $P(E)$, and so none for supposing the sample is misleading.

This bears on another matter. Suppose an experimenter completes his experiment, and infers from the sample that the best supported hypothesis about the proportion of $E$'s is that it is $p$. Someone else notices that every member of his sample has some property $B$ which no one would be inclined to account a 'trick' property like the one I just described. There is still no ground for criticism of the experimental result. It is of course open to the sceptic to try another sample in which not all members are $B$'s, but unless there is good reason to suppose, for instance, that $P(E)$ differs from $P(E|B)$, it remains the case that on the data the best supported hypothesis about the proportion of $E$'s is $p$.

*Random sampling numbers*

In practice it is very hard to make a set-up, experiments on which are independent. The obvious solution is to have a standard chance set-up and anchor all random sampling to it. If every member of a population of interest can be labelled by a serial number, the problem reduces to sampling serial numbers. Tables of random sampling numbers—random numbers, for short—are now printed, and these should provide random samples from serial numbers. From their humble beginnings in Tippett's tables, published in 1927, these tables have swollen to RAND's million random digits, issued in 1955. Just to emphasize the difficulty of producing a set-up on which trials are independent, even the electronic RAND machine, seemingly as perfect a thing as man could devise, exhibited dependencies and marked bias just like earlier and much less sophisticated apparatus. The dependencies were tracked down to mechanical faults, so these negative results do not show there could never be a truly independent machine, but only that this machine showed slight dependencies.

Since the notion of randomness can produce, in minds unfamiliar with laboratory practice in designing independent experiments, either a feeling of mystification or else a positive mystique of randomness, it may be useful to point out what sort of mechanical effects can be expected in random number machines. A machine producing a host of digits has to work pretty fast or else it would never be finished in time for impatient humans. Electronic science is a suggestive source of speedy random processes. 'Shot noise'— the noise picked up in ear phones and which is produced by the random behaviour of swarms of electrons and which sounds like dropping masses of lead shot in a tin plate—was one of the earliest sources of randomness provided by physics. However, an electric circuit is needed to measure variations in shot noise and, indeed, to amplify it to the point where it can be measured. Such circuits are generally equipped with condensers. Condensers store charges, and release them gradually. The decay time, or the rate at which a condenser releases a charge, is a property of the condenser and the charge. The properties of a circuit to some degree depend on the charges stored on the condensers. Hence circuits have memories: their properties at one instant of time are to some extent a function of their properties at preceding instants. Even if electrons do not have memories, circuits measuring them do. Even if the memories are short, a machine producing lots of digits quickly has to work and count in periods shorter than typical decay times. Hence even if the behaviour of electrons does itself produce a chance set-up on which trials are independent, measurement of the results of those trials may be systematically inaccurate, and build up a systematic bias. This is part of the reason why even the most recent work on electronic production of random numbers has not been entirely satisfactory—not much more satisfactory, indeed, than work based on counting dates in eighteenth-century parish registers. This is a fact about physics, not about the logic of randomness.

*Use of random numbers*

You might explain a table of random sampling numbers as a sequence of numbers produced on consecutive trials on a device in which every digit has equal chance at every trial. But this is only a rough approximation. For such a table is not something to be admired in a random number gallery but a tool for the design of experiments. That is why I use the old-fashioned expression 'random sampling numbers' rather than the briefer 'random numbers'; the original term makes plain that these are digits for random sampling.

It might seem that an ideal table of random sampling numbers is one such that, by starting at any point in the sequence of digits, one could use the table to select any sample of serial numbers of any size, and proceed to normal statistical inference assuming a random sample. But this ideal is notoriously unobtainable. One table begins by saying 'A set of Random Sampling Numbers should be of such a kind as to give results approximately in accordance with expectation when used in moderately small amounts or en bloc'.†
But as was realized at once

it is impossible to construct a table of random sampling numbers which will satisfy this requirement entirely. Suppose, to take an extreme case, we construct a table of $10^{10^{10}}$ digits. The chance of any digit being a zero is $1/10$ and thus the chance that any block of a million digits are all zeros is $10^{-10^6}$. Such a set should therefore arise fairly often in the set of $10^{10^{10}-6}$

† M. G. Kendall and B. B. Smith, *Tables of Random Sampling Numbers* (Cambridge, 1939), p. viii.

USE OF RANDOM NUMBERS 131

blocks of a million. If it did not, the whole set would not be satisfactory for certain sampling experiments. Clearly, however, the set of a million zeros is not suitable for drawing samples in an experiment requiring less than a million digits.

Thus, it is to be expected that in a table of random sampling numbers there will occur patches which are not suitable for use by themselves. The unusual must be given a chance of occurring in its due proportion, however small. Kendall and Babington Smith attempted to deal with this problem by indicating the portions of their table (5 thousands out of 100) which it would be better to avoid in sampling requiring fewer than 1000 digits.†

I think the solution of Kendall and Babington Smith is correct but there is a defect in this account of the matter. The authors say that in one circumstance, the whole set of numbers would not be satisfactory for certain experiments and that, in the converse circumstance, some parts of the set would not be suitable for other experiments. Two entirely different things are at stake here. One source of dissatisfaction is due to lack of randomness, the other to a random sample's being misleading.

First, if a device could produce no blocks of a million zeros, it would be unsatisfactory because trials on it would not be independent. If an enormous number of digits contained no such blocks, the digits themselves would give good evidence that trials on the device were not independent. Hence one could not use these digits for random sampling, for the premise of random sampling, that trials are independent, would itself be rejected. Thus one kind of unsatisfactoriness stems from trials not being random.

But on the other hand suppose there is no indication that the trials are not random, and so that there are occasional blocks of a million zeros. These blocks will not be suitable for small trials not because sampling will not be random, but because the sample will be misleading. In the extreme case, if you wanted to sample 100 members from some population and started in the middle of a block of a million zeros, you would single out the same individual over and over again. If there were reason to believe that the proportion of $E$'s in the population as a whole differs from 0 or 1, then your sample used to discover the proportion of $E$'s in the

† M. G. Kendall and A. Stuart, *Advanced Theory of Statistics* (3 vol. ed.) (London, 1958), 1, pp. 217 f.

population would necessarily be misleading in the sense explained in the preceding section.

In this connexion, it is worth noting that quite aside from having parts of a table of random numbers useless for small samples, a particular experiment using random numbers may be rejected for similar reasons. To take the sort of consideration which much influenced Fisher, suppose in agricultural trials that seeds of different varieties are to be planted in plots on a field. It is known that not all plots are equally fertile. In particular, suppose fertility increases as you go from the north-east corner to the south-west one. Then in testing seeds of different sorts, the variety to be planted in each plot might be settled by use of a table of random numbers. Now suppose that on some occasion your random numbers tell you to plant all the seeds of one variety in the north-east half of the field, and the rest in the other part. It is obviously correct not to use this arrangement, because the result would be misleading. That is, one hypothesis best supported on the data about random allotment and actual growth would not be best supported on the fuller data about the fertility gradient in the field.

The fact that random sampling numbers must be used with tact follows of necessity from the general theory of statistical support. It is perhaps especially noteworthy that in all these investigations, there has been no mention of the concept of a random sequence. Overemphasis on that probably useless concept can lead to grievous error. The author of a recently published budget of paradoxes is at great pains to prove the truism that any sequence of digits satisfactory for use in random sampling on a very long sequence of trials will contain stretches which are not suitable for use on samples with a fairly small number of members. This he takes to show up 'the ultimate self-contradiction of the concept of randomness'.† At most it shows that the concept of random sequence is not of any use. Fortunately no serious statistician has ever relied on it.

† G. Spencer Brown, *Probability and Scientific Inference* (London, 1957), p. 57.

CHAPTER IX

# THE FIDUCIAL ARGUMENT

Much can be done with mere comparisons of support, but statisticians want numerical measures of the degree to which data support hypotheses. Preceding chapters show how much can be achieved by an entirely comparative study. Now I shall argue that our earlier analysis can sometimes provide unique quantitative measures. It is not yet certain that this conclusion is correct, for it requires a new postulate. However, this is interesting in its own right, and though the resulting measures will have a fairly narrow domain, their very existence is remarkable.

The following development is novel, but in order to declare its origins I call this chapter *the fiducial argument*. The term is Fisher's. No branch of statistical writing is more mystifying than that which bears on what he calls the fiducial probabilities reached by the fiducial argument. Apparently the fiducial probability of an hypothesis, given some data, is the degree of trust you can place in the hypothesis if you possess only the given data. So we can at least be sure that it is close to our main concern, the degree to which data support hypotheses.

Fisher gave no general instruction for computing his fiducial probabilities. He preferred to work through a few attractive examples and then to invite his readers to perceive the underlying principles.† Yet what seem to be his principles lead direct to contradiction.

Despite this bleak prospect, the positive part of the present chapter owes much to Fisher, and can even claim to be an explication of his ideas. Fortunately the consistent core of his argument is a good deal simpler than his elliptic papers have suggested. There has already been one succinct statement of it: Jeffreys was able not only to describe its logic, but also to indicate the sort of postulate

† The original paper is 'Inverse probability', *Proceedings of the Cambridge Philosophical Society*, XXVI (1930), 528–35. Perhaps the clearest statement of his views is in 'The logical inversion of the notion of a random variable', *Sankhya*, VII (1945), 129–32. Accounts of the matter recur in *Statistical Methods and Scientific Inference* (Edinburgh, 1956).

Fisher must assume for his theory to work.† It is a shame that Jeffreys' insights have not been exploited until now.

## Logic

If this chapter does contribute either to understanding Fisher's principles or to the sound foundation of a theory of quantitative support, it will only be through careful statement of underlying assumptions and conventions. It is by no means certain what ought to be the basic logic of quantitative support by data. There is, however, a pretty wide concensus that the logical axioms should coincide with those of the probability calculus, that is, with Kolmogoroff's axioms. I should not myself care to assume this with any generality, but we shall accept those axioms for the special case of support for statistical hypotheses by statistical data.

There are really two questions about the axioms, one a matter of fact, and one a matter of convention. It is a question of convention to choose manageable axioms which are consistent with the principles. Jeffreys has shown how great is the element of convention in the probability axioms taken as postulates for measuring support: he argues that even additivity is conventional. But a more recent study shows that any practicable set of axioms must follow the traditional course.‡

For our purposes we take a Boolean algebra or sigma-algebra of statistical hypotheses, assertions about particular trials on chance set-ups, and other propositions built up from these by conjunction, alternation, and negation, and, in the case of a sigma-algebra, countable alternation. Hitherto we have used a capital $P$ for chance; now let us use a small $p$ for support. We state the following axioms, with '$p(h|e)$' interpreted as the degree to which $e$ supports $h$.

(1) $0 \leqslant p(h|e) \leqslant 1$.
(2) $p(h|h) = 1$.
(3) If $h$ and $i$ are logically equivalent, $p(h|e) = p(i|e)$ and $p(e|h) = p(e|i)$.
(4) $p(h \& i|e) = p(h|e \& i) p(i|e)$.
(5) If $h$ and $i$ are incompatible, $p(h|e) + p(i|e) = p(h \vee i \,|e)$.

In the case of a Boolean sigma algebra, when we may have countable alternations, we must extend the final axiom. It will be

---

† H. Jeffreys, *Theory of Probability*, p. 301.
‡ R. T. Cox, *The Algebra of Probable Inference* (Baltimore, 1961), ch. 1.

recalled that $h$ is a countable alternation of a countable set of propositions $\{h_n\}, (n = 1, 2, \ldots)$, just when $h$ is true if and only if some $h_n$ is true. We extend (5) to,

(5*) If $h$ is a countable alternation of the set of mutually exclusive propositions $\{h_n\}, (n = 1, 2, \ldots)$ then $p(h|e) = \Sigma p(h_n|e)$.

Evidently these axioms are included in the Kolmogoroff system with $P(\ |\ )$ interpreted as $p(\ |\ )$, while $P(\ )$ is assigned no interpretation. Or else we can take $P(\ )$ as $p(\ |t)$, where $t$ is a necessary truth. Hence from a formal point of view there is nothing new in our axioms for $p$.

## The frequency principle

One principle about support and chance seems so universally to be accepted that it is hardly ever stated. It is not so much a principle as a convention to which everyone is firmly wedded. Roughly, if all we know are chances, then the known chances measure degrees of support. That is, if all we know is that the chance of $E$ on trials of kind $K$ is $p$, then our knowledge supports to degree $p$ the proposition that $E$ will occur on some designated trial of kind $K$.

More exactly, let $d$ represent the data that on trials of kind $K$ the chance of $E$ is $p$. Let $h$ be the proposition that $E$ occurs on a designated trial of kind $K$. Then it seems universally to be agreed that the number $p$ is a good measure of the degree to which $d$, taken alone, supports $h$. Use of $p$ may be conventional; perhaps other functions of $p$ would do. It is convenient to take the simplest function. Given the axioms for chance and degrees of support, $p$ itself is incomparably more simple than any other function of $p$. Our convention is not entirely empty. In the terminology of joint propositions, we have, first of all, that if $P_D(E)$—the chance of $E$ according to $D$—exceeds o, then,

$$p(\langle X, K, D; T, K, E \rangle | \langle X, K, D; T, K, \Omega \rangle) = P_D(E).$$

More generally suppose trials of kind $K'$ are derivative on trials of kind $K$, and that whenever the distribution of chances on trials of kind $K$ is in the class $\Delta$, then the chance of $E$ on a trial of kind $K'$ is some definite value $P_\Delta(E)$, greater than o. Then,

$$p(\langle X, K, \Delta; T, K', E \rangle | \langle X, K, \Delta; T, K', \Omega \rangle) = P_\Delta(E).$$

This assertion will be called the *frequency principle*.

This complex statement should not obscure the triviality of the principle. Knowing only that the chance of drawing a red ball from an urn is 0·95, everyone agrees, in accordance with the law of likelihood, that a guess of 'red' about some trial is much better supported than one of 'not-red'. But nearly everyone will go further, and agree that 0·95 is a good measure of the degree to which 'red' is supported by the limited data. Indeed, a weak form of the frequency principles can be derived from the probability axioms for support, plus the law of likelihood.

It is widely held that the single word 'probability' denotes both long run frequency and what I call the degree to which one proposition supports another. If this were true of some dialect, then in that dialect the frequency principle could be expressed in a very boring way. The probability is the probability. But our principle is a little more substantial than that. Avoidance of the word 'probability' will pay dividends in this chapter, for many discussions of the fiducial argument have foundered on what that word means. However, I might suggest that the meaning of the English word 'probability' is very similar to that of 'degree of support'; it has become entwined with long run frequency just because, in certain standard and influential cases of limited data, long run frequencies provide the natural measure of degrees of support.

*A trivial example*

The fiducial argument has collected so many statistical barnacles that it is worth beginning on dry land with no hint of statistical inference. Imagine a box and an urn. The box contains a coloured ball. You want to know its colour but cannot open the box. You are told that the urn contains 100 balls; 95 of the balls are the same colour as that in the box. The chance of drawing a ball of some colour is exactly in proportion to the number of balls of that colour in the urn. Hence the chance of drawing a ball the same colour as the boxed ball must be 0·95. Suppose one ball is drawn, and it is red. Everyone agrees it is a good bet, that the boxed ball is also red. Indeed 0·95 seems to measure the degree to which the whole set of data supports the hypothesis, that the ball in the box is red. Can this be justified?

First let data $d$ be our initial data, stating all our knowledge except that the ball drawn is red. $e$ shall state this final bit of

## A TRIVIAL EXAMPLE

information, that a red ball is drawn. Now consider the hypothesis,

*h*: The ball drawn from the urn is the same colour as the ball in the box.

By the frequency principle, we may argue that *d* supports *h* to degree 0·95

$$p(h|d) = 0·95. \qquad (1)$$

Now *e* states that a red ball is drawn from the urn. *If* we can accept that when only *d* is known the redness of the ball counts neither for nor against *h*; *if* we can accept that the redness is simply irrelevant to *h* in the light of *d*; *then* we could conclude that *e* makes no difference to the support for *h*. That is, (1) would imply

$$p(h|d\&e) = 0·95. \qquad (2)$$

Now consider the hypothesis,

*i*: The ball in the box is red.

*d*&*e* implies that *h* is equivalent to *i*, or, as I shall say, that *i* and *h* are equivalent, given *d*&*e*. Hence the logic of support entails that

$$p(h|d\&e) = p(i|d\&e). \qquad (3)$$

(2) and (3) together imply that $p(i|d\&e) = 0·95$, which was the conclusion suggested in the first paragraph of this section. Everything hinges on the single premise of irrelevance needed to get from (1) to (2). What Fisher called the fiducial argument can be made to work in exactly the same way as this trifling example, and it needs a premiss of exactly the same sort. Irrelevance will be the lynch pin of this chapter.

### *A statistical example*

Now we run through the same course of argument in the most elementary statistical example. Once again we shall not consider whether an assumption of irrelevance is true, but only show how it leads to an interesting inference. We take a coin tossing experiment. It is true that Fisher explicitly denied that his argument could work where there are only finitely many possible results, but he probably discarded the following sort of case as too insignificant. Yet this case is uncommonly useful in making plain the logic of his method.

Let data *d* state that the chance of heads on independent trials on a coin tossing device is either 0·6 or 0·4, and that another toss is to be made. There is a roundabout way of evaluating the significance of $d$ to the two possible hypotheses about the chance of heads. Tossing the coin constitutes a primary kind of trial. Many derivative kinds of trial can be defined in terms of it. Here is one, a little bizarre, but with an important property. On the derivative kind of trial let there be two possible results, labelled 0 and 1. Which of these occurs on any particular derivative trial shall depend on the outcome of the corresponding primary trial, according to the following table:

0 occurs if $\begin{cases} \text{heads occurs and the chance of heads is 0·6, or} \\ \text{tails occurs and the chance of heads is 0·4.} \end{cases}$

1 occurs if $\begin{cases} \text{heads occurs and the chance of heads is 0·4, or} \\ \text{tails occurs and the chance of heads is 0·6.} \end{cases}$

In the language of random variables, we have defined a new random variable, with possible values 0 and 1, in terms of an old random variable, with possible values heads and tails.

The newly defined derivative kind of trial has the striking property that given data $d$, the chances of 0 and 1 are known. For if the chance of heads were 0·6, 0 would occur only if heads came up, and so the chance of 0 would be 0·6; but if the chance of heads were 0·4, 0 would occur only if tails came up, and once again the chance of 0 would be 0·6. Either way the chance of 0 is 0·6, and likewise the chance of 1 is 0·4. So $d$ tells us the distribution of chances on our derivative kind of trial.

Hence we can use the frequency principle, just as with the box and urn above. Since $d$ tells us the chance of 0 on trials of the derivative kind, we have from the frequency principle that, if $T$ stands for the next derivative trial,

$$p(\text{0 occurs at } T | d) = 0·6. \tag{4}$$

So much relies only on the frequency principle. Now we conduct trial $T$, and observe tails, say; $e$ shall be the proposition stating that tails happens at $T$. It may seem that $e$—occurrence of tails—is simply irrelevant to whether or not 0 has occurred. This may appear from considerations of symmetry in the definition of 0 and 1. Or you may reflect that occurrence of 0 means that either heads or tails occurred, and whichever it was, the odds were in favour of what actually happened; occurrence of 1 means that either heads or tails occurred, and that the odds were against what

A STATISTICAL EXAMPLE 139

happened. And so it may look as if *e* is irrelevant to whether or not o occurred. Learning that tails happened is no help at all to guessing whether o happened. I shall later give different and sounder reasons for supposing that *e* is irrelevant to occurrence of o. For the present we note merely that *if e* were irrelevant, given *d*, to whether or not o occurred at the designated trial *T*, we should have from (4) that,

$$p(\text{o occurs on trial } T | d \& e) = 0 \cdot 6. \tag{5}$$

But as o is defined, occurrence of tails implies that o occurs if and only if $P(H) = 0 \cdot 4$. Hence by the logic of support (5) implies,

$$p(P(H) = 0 \cdot 4 | d \& e) = 0 \cdot 6. \tag{6}$$

(6) is remarkable. It states the degree to which the statistical data support a statistical hypothesis. This is exactly what we are looking for: a measure of the degree to which data support hypotheses. We have got it, at least in this special case, from the feeble frequency principle plus some assumption about irrelevance. So now we must examine whether this, or another, assumption of irrelevance could ever be justified.

*The structure of the argument*

Before studying justifications let me set out again, on pain of redundancy, the structure of the argument.

(1) Begin with some initial data *d* about trials of kind *K* on a set-up.

(2) Define a derivative kind of trial, say *K'*, such that, even if the distribution on trials of kind *K* is unknown, *d* does imply a unique distribution of chances of outcomes on trials of kind *K'*.

(3) Use the frequency principle to get the degree to which *d* supports various hypotheses about the outcome of a designated trial of kind *K'*.

(4) Consider further data *e*.

(5*) Suppose that given *d*, *e* is irrelevant to hypotheses about the outcome of a designated trial of kind *K'*.

(6) From (3) and supposition (5*) infer the degree to which *d*&*e* supports various hypotheses about the outcome of the trial of kind *K'*.

(7) Observe that given *d*&*e* hypotheses about the outcome of the trial of kind *K'* are equivalent to hypotheses about the distribution of chances of outcomes on trials of kind *K*.

(8) Infer from (6) and (7) the degree to which $d \& e$ supports hypotheses about the distribution on trials of kind $K$.

Evidently the structure of the argument is entirely general. The step marked with a star shows where we need a further premiss. The other novelty is at step (2), and it will be useful to have a name for derivative trials with the property required for (2). Given a set-up, trials of kind $K$, and data $d$, trials derivative on trials of kind $K$ shall be called *pivotal* if the distribution of chances on these derivative trials is uniquely determined by $d$.

In our example, the result of our derivative trial is a function of the result of the corresponding primary trial, plus the true chances on trials of kind $K$. When the function determining a derivative kind of trial determines a pivotal kind of trial, then I shall say that the function itself is *pivotal*. Finally, and only here do I pick up Fisher's terminology, the result of a pivotal kind of trial can be referred to as a *pivotal quantity*. The word 'pivotal' has been used chiefly to preserve some kind of historical continuity. Part of Fisher's original idea was to interchange some of the variables in the distribution for the pivotal kind of trial, and so in a sense to pivot on the relevant equations. But I am afraid the etiology of the term does not bear close scrutiny. Fisher did not use it at first, and I am not sure who invented it.

If, given trials of kind $K$ and data about them, we can construct a pivotal function and a pivotal kind of trial, we have got to step (2), and can proceed along to (5*), where we suppose irrelevance. But of course not all suppositions about irrelevance need be true. Some may contradict others. The task of this chapter is to define which assumptions are true, or at least to conjecture a principle of irrelevance, and to begin the task of testing its truth.

Fisher invented the fiducial argument. At any rate, I suppose that what he called the fiducial argument must, if it has any validity, take something like the course described above. Taking liberties with the exact text of his writings, it seems that he was content to say that some pivotal quantities are 'natural' while others are not. He was happy to believe that 'we can recognize' that, in effect, some data are irrelevant to some hypotheses. No one has yet been able to state exactly what we, or Fisher, are supposed to recognize. Only Jeffreys could indicate exactly where a principle of irrelevance is needed, but even he did not explore the difficulties. So it remains a conceptual problem to analyse irrelevance. It would be important to do so if no contradictions were known in connexion with the fiducial argument; it is essential since tactless application of the argument spawns myriad contradictions.

If you suppose, as Fisher once did, that given a pivotal function you can always assume irrelevance at step 5*, you get paradoxes. There are too many pivotal functions in connexion with most important problems, and it turns out to be impossible to maintain that the result of any pivotal kind of trial is irrelevant to the statistical hypotheses of interest. A technical note toward the end of the chapter will describe some of the known contradictions, but first let us do some more positive work, and discover a consistent principle which says when we can make step (5*), and when we cannot.

*Irrelevance*

The idea of the forthcoming principle of irrelevance follows directly from the analysis of the earlier chapters. Relative support has been analysed in terms of relative likelihoods. If new evidence $e$ is irrelevant, given the old data $d$, to some exclusive and exhaustive set of propositions, then the relative support furnished for the different propositions by $d$ should be the same as that furnished by $d \& e$. And if relative support is as relative likelihood, this should occur just if the likelihoods given $d \& e$ are the same as given $d$.

Suppose for example that we entertain as hypotheses the exclusive and exhaustive $h$ and $i$, and that joint proposition $d$ conveys all we know about their truth. Let $d$ support $h$ better than $i$. Suppose we subsequently learn the further facts $e$, so that we become possessed of $d \& e$. Naturally, $h$ might no longer be better supported than $i$; $d \& e$ might favour $i$ instead. But there is at least one situation when, according to the law of likelihood, there will be no change at all: if the likelihood of $h$, given $d$, is the same as given $d \& e$, and likewise for $i$. We should have learned nothing from $e$ about $h$ as opposed to $i$. In this special situation, $e$ is simply irrelevant to $h$ and $i$. I shall further conjecture that the ratio between the support $d$ furnishes for $h$ and the support $d$ furnishes for $i$, is the same as the ratio when $d \& e$ is the datum.

This conjecture will be formalized as the *principle of irrelevance*. I cannot see it as a new departure, but only as the overt statement of something implicit in our earlier analysis. But like any other conjecture, though it may gain some plausibility from heuristic considerations it is primarily to be tested by its consequences.

To apply the principle to the coin tossing example, we would have two mutually exclusive hypotheses: that o occurred at trial $T$, and that 1 occurred. We have to show that the likelihood that o occurred at trial $T$, given $d$, is the same as given $d \& e$; likewise for 1. I shall work this through in the next section. First let us review the technical forms essential to a precise exposition: the reader may prefer to omit this, and proceed to the next section immediately, referring to this review only to recall the meaning of particular terms.

A *simple joint proposition* is a conjunction of two propositions about a set-up: it is a conjunction of (*a*) an hypothesis stating the distribution of chances of outcomes on trials of some kind, and (*b*) an hypothesis stating the outcome which occurred on some particular trial of that kind. The likelihood of a simple joint proposition is what the chance of outcome (*b*) would be, if the distribution (*a*) were correct. Such likelihoods are 'absolute' likelihoods, being no more than numbers assigned to simple joint propositions in accordance with the two conjuncts (*a*) and (*b*).

The more general notion of *joint proposition* is also defined as a conjunction, this time of (*c*) an hypothesis that the distribution of chances on trials of some kind on a set-up does fall in some stated class of distributions, and (*d*) an hypothesis stating the outcome of a trial of some (possibly different) kind on the same set-up. We represented joint propositions by the symbol $\langle X, K, \Delta; T, K', E \rangle$—set-up $X$, kind of trial $K$, class of distributions $\Delta$; trial $T$ of kind $K'$ with outcome $E$. If $D$ is a unique distribution, a simple joint proposition would be of the form $\langle X, K, D; T, K, E \rangle$. We say that such a proposition is *included in* a joint proposition $e$ if $e$ is logically equivalent to a joint proposition $\langle X, K, \Delta; T, K, E' \rangle$, and $D$ is a member of $\Delta$, while $E$ is contained in $E'$.

The *likelihood of h given d* was defined in two stages. First, for simple joint propositions $h$, the likelihood of $h$ given any $d$ which includes $h$ is just the 'absolute' likelihood of $h$. Secondly an equivalence condition. If $d$ implies that propositions $h$ and $i$ are equivalent, so that $h$ and $i$ are equivalent given $d$, then the likelihoods of $h$ and $i$ given $d$ are the same.

The *law of likelihood*, so essential for sound statistical inference, was expressed in two different ways. The first, and so far most often used, is in terms of absolute likelihoods, while the second uses likelihoods given data. The two versions are equivalent within the logic of comparative support. The first version says that if joint proposition $d$ includes the simple joint propositions $h$ and $i$, while the absolute likelihood of $h$ exceeds that of $i$, then $d$ supports $h$ better than $i$. In the second way of putting the matter, we say *d supports h better than i if the likelihood of h given d exceeds the likelihood of i given d*.

## Application to the statistical example

As a preface to a rigorous statement of the principle of irrelevance, let us see how it works on the elementary statistical example. We had data $d$ stating,

$d$: The chance of heads on independent trials of kind $K$ is either 0·6 or 0·4; and trial $T$ of kind $K$ occurs, with result either heads or tails.

We further defined a derivative kind of trial, call it of kind $K'$. Corresponding to the trial $T$ of kind $K$ is the trial $T'$ of kind $K'$. On any trial of kind $K'$, 0 occurs just if heads occurs on the corresponding trial of kind $K$, and the chance of heads is 0·6, or tails occurs on the corresponding trial, and the chance of heads is 0·4. Hence $d$ is equivalent to another joint proposition $d'$:

$d'$: The chance of heads on independent trials of kind $K$ is either 0·6 or 0·4; and trial $T'$ of kind $K'$ occurs with result either 0 or 1.

In our more abbreviated notation, we could write:

$d$: $\langle X, K, P(H)$ is 0·4 or 0·6; $T, K$, heads or tails$\rangle$,
$d'$: $\langle X, K, P(H)$ is 0·4 or 0·6; $T', K'$, 0 or 1$\rangle$.

$d'$ can be split into two halves, keeping the first conjunct intact, but dividing the second, as follows,

(1) $\begin{cases} (1a) \ \langle X, K, P(H) \text{ is 0·4 or 0·6}; T', K', 0\rangle, \\ (1b) \ \langle X, K, P(H) \text{ is 0·4 or 0·6}; T', K', 1\rangle. \end{cases}$

This division of $d'$ into mutually exclusive propositions will be called a *partition* of $d'$. Since we know the chance of 0 according to $d$, or according to the equivalent $d'$, the frequency principle gives the degree to which $d$ (or $d'$) supports each of (1a) and (1b). The chance of 0 is 0·6 and the chance of 1 is 0·4. Hence $d'$ supports (1a) to degree 0·6, and (1b) to degree 0·4. Likewise for $d$.

Now let $e$ stand for the further piece of information, that tails actually occurs on trial $T$. $d\&e$ is simply,

$d\&e$: $\langle X, K, P(H)$ is 0·4 or 0·6; $T, K$, tails$\rangle$.

We cannot tell directly how well $d\&e$ supports (1a). So first of all, take a partition of $d\&e$, this time splitting up the first conjunct as follows:

(2) $\begin{cases}(2a) & \langle X, K, P(H) \text{ is 0·4}; T, K, \text{tails}\rangle, \\ (2b) & \langle X, K, P(H) \text{ is 0·6}; T, K, \text{tails}\rangle.\end{cases}$

(1a) is certainly included in $d'$, and hence in the equivalent $d$. So its likelihood given $d$ is just its absolute likelihood, namely the chance of 0. As this chance was shown to be 0·6, we have,

The likelihood of (1a), given $d$, is 0·6,

Similarly,

The likelihood of (1b), given $d$, is 0·4.

In a parallel manner, we observe that (2a) and (2b) are included in $d\&e$, and compute that,

The likelihood of (2a), given $d\&e$, is 0·6,
The likelihood of (2b), given $d\&e$, is 0·4.

It will also be observed that, given $d\&e$, (1a) is equivalent to (2a), and (1b) to (2b). So by our second stipulation for computing likelihoods given data, we infer from the last stated pair of likelihoods that,

The likelihood of (1a), given $d\&e$, is 0·6.
The likelihood of (1b), given $d\&e$, is 0·4.

The first and third pairs of likelihoods show that the likelihood of (1a) given $d$ is the same as that given $d\&e$, and likewise for (1b). So on our suggested principle of irrelevance, $e$ is simply irrelevant to (1a), given data $d$. We know the degrees to which $d$ supports (1a) and (1b); 0·6 and 0·4 respectively. So our suggested principle of irrelevance implies,

$p((1a)|d\&e) = 0\cdot 6,$
$p((1b)|d\&e) = 0\cdot 4.$

Mere manipulations of the logic of support, relying on the fact

## APPLICATION TO THE STATISTICAL EXAMPLE

that $d\&e$ implies the equivalence of (1 $a$) and the proposition that $P(H) = 0.4$, suffice to conclude that,

$$p(P(H) = 0.4|d\&e) = 0.6,$$
$$p(P(II) = 0.6|d\&e) = 0.4.$$

As predicted, a principle of irrelevance based on likelihoods is going to provide a measure of the degree to which some data support some hypotheses.

*Terminology*

A few purely logical terms simplify a precise statement of the proposed principle of irrelevance. As already indicated in the course of the example, a *partition* of $e$ is a set of mutually exclusive propositions such that $e$ implies some member of the set is true. (It is not demanded that the truth of any member should imply $e$.) A partition of $e$ into simple joint propositions is a partition of $e$, all of whose members are simple joint propositions. *Negation with respect to $e$* is the operation which forms $(\sim h)\&e$ from $h$. In set theory the corresponding notion is complementation within some universe.

Finally, a *logical field* based on a partition $A$ of $e$ is the smallest set of propositions closed under the operations of countable alternation and negation with respect to $e$, and which includes every member of the partition $A$.

When there are only finitely many possible results for trials of some kind, the principle of irrelevance can be stated easily. First, if $A$ is a partition of both $d$ and $d\&e$, while the likelihood of each member of $A$, given $d$, equals its likelihood given $d\&e$, then $e$ is said to be *irrelevant* to $A$, given $d$. Our principle is: *if $e$ is irrelevant to $A$ given $d$, and $h$ is in the logical field based on $A$, then*

$$p(h|d) = p(h|d\&e).$$

In our trivial example, the partition $A$ in question consisted just of (1 $a$) and (1 $b$). We showed that $e$ was irrelevant to this partition, given $d$, and were then able to infer, first that

$$p((1a)|d\&e) = p((1a)|d) = 0.6,$$

and then, by mere logic, that

$$p(P(H) = 0.4|d\&e) = 0.6.$$

*Initial support*

Before passing to more exciting set-ups with continuously many possible results, it will be useful to extract a moral from the finite case. We began with data $d$, and augmented it by data $e$. Using the principle of irrelevance we inferred the degree to which $d \& e$ supports various hypotheses. But we did not discover the degree to which $d$, taken alone, supports hypotheses about the chance of heads. I can see no reason to suppose that in general such a degree of support should even exist. But if it does exist, let us call it the initial support, the support provided by initial data $d$. If it does exist, the initial support $p(h|d)$ is subject to the following consequence of the probability axioms, discovered long ago by Baves:

$$p(h|d \& e) = \frac{p(e|h \& d) p(h|d)}{p(e|h \& d) p(h|d) + p(e| \sim h \& d) p(\sim h|d)}.$$

Here let $d$ and $e$ be as before, and $h$ the hypothesis that the chance of heads is 0·4. We have computed $p(h|d \& e)$ via the principle of irrelevance; routine manipulations enable us to solve the equation for $p(h|d)$; we have

$$p(h|d) = p(\sim h|d) = 1/2.$$

This is a useful formula. In our original problem data $e$ covered only the result of a single toss. We could not directly apply the fiducial argument to the result of two tosses. But since the above argument gives the initial support which $d$ furnishes for $h$ and $\sim h$, we can, by another use of Bayes' formula, compute the degree to which $d \& e'$ supports $h$, where $e'$ is the result of any sequence of trials whatsoever.

The importance of the initial distribution should not be overestimated. We could have proceeded without it. For we know the degrees of support furnished by $d \& e$, where $e$ is the result of the first toss. If $e'$ is the result of further trials, we may write Bayes' formula as

$$p(h|d \& e \& e') = \frac{p(e'|h \& d \& e) p(h|d \& e)}{p(e'|h \& d \& e) p(h|d \& e) + p(e'| \sim h \& d \& e) p(\sim h|d \& e)}$$

and compute the desired degree of support.

## Indifference principles

$P(H) = 0.6$ and $P(H) = 0.4$ were, according to $d$, the only possible hypotheses; each is supported by $d$ to degree $1/2$. This is reminiscent of the notorious principle of indifference, which states that if a body of data supplies a finite number of alternatives, and provides no further information, then it supports each alternative equally well. Here we need only recall that this noxious principle has never been stated so as to provide a consistent measure of support which is widely applicable. The trouble is seen easily enough. $d$ mentions only two hypotheses, but it equally 'supplies' the following three: (1) $P(H) = 0.6$, and there is life on some celestial body other than earth, (2) $P(H) = 0.6$ and there is no such life, and (3) $P(H) = 0.4$. Are we to take $1/3$ as the degree to which alternative (3) is supported?

It is true that there is a residual, if unformalizable, intuition that if the chance of heads must be either $0.6$ or $0.4$, then, lacking other information, each of the alternatives is equally well supported by the feeble data available. So it is pleasant that our theory should not conflict with this tempting alleged intuition. But mere lack of conflict with something so nebulous cannot be taken as evidence that our foundations are correct.

A question remains. It could be said that the principle of irrelevance merely expresses a judgement about irrelevance. From this judgement follows a distribution of initial support. But from a mathematical point of view, one could equally well begin with some judgement about initial support and derive the claim about irrelevance. This, if I understand it, is Jeffreys' view of the matter.†
I shall argue that it is not correct.

The principle of irrelevance has been stated in complete generality. It makes no mention of the special conditions of any problem. Its central notion, that degrees of support are as relative likelihoods, originates in the simplest conceptions about frequency. It seems merely to continue that explication of the fundamental fact about frequency and support which led to the laws of likelihood. And even then the principle of irrelevance does not produce numerical degrees of support *de novo*, but merely, as it were,

† H. Jeffreys, *Theory of Probability*, p. 301.

computes new probabilities out of those already established by the conventional frequency principle.

Thus if the principle of irrelevance is true at all, it seems to express a fundamental fact about frequency and support. But turning to alleged judgements of indifference, we find no fundamental fact to express. In this or that problem one can conceive some distribution of support which more or less conveniently represents ignorance or initial data. But it is impossible to state any consistent rules as general as the principle of irrelevance. At the very most one gets rules which span all the cases where the parameters have some designated range. In particular problems there are *ad hoc* judgements of indifference which coincide with the measures of support provided by the frequency principle and the principle of irrelevance. From a mathematical point of view it may not matter much, except in point of simplicity, which we begin with, a supposedly 'correct' and entirely *ad hoc* judgement of indifference, or with the principle of irrelevance. But from a philosophical point of view only one can appear fundamental.

However, all this is mere heuristic argument directed at the case of finitely many possible results. In that case, whenever the fiducial argument works, we could always have begun instead with a judgement of indifference. The matter is not definitely settled. It is settled when we turn to the continuous case where a much more interesting situation develops: a striking contrast between the principle of indifference approach, and that based on the principle of irrelevance. The contrast heavily favours the latter principle with its associated fiducial argument.

*Likelihood functions*

To get into the case of continuously many possible results, the principle of irrelevance will have to be expressed in terms of likelihood ratios rather than likelihoods. It will be recalled that when the likelihoods of both $\langle X, K, D; T, K, E \rangle$ and $\langle X, K, D'; T, K, E' \rangle$ are greater than zero, their likelihood ratio is the ratio of their likelihoods. When, in the continuous case, $E$ and $E'$ are both results, the likelihood ratio of the two simple joint propositions is the ratio of the experimental density of $D$ at $E$ to that of $D'$ at $E'$.

Experimental densities were defined on pp. 68–70 above, but here is a brief review. If results of trials of kind $K$ are of the form

$z_1, z_2, ..., z_n$, an ordered set of real numbers, the cumulative distribution function for any $D$ will be of the form $F(z_1, z_2, ..., z_n)$. If the values of $z_1, z_2, ..., z_n$, at a trial of kind $K$ are determined by measurements on the quantities $x_1, x_2, ..., x_m$, which may vary from trial to trial, then the experimental density function is the partial derivative of $F$ with respect to the quantities $x_1, x_2, ..., x_m$, and expressed as a function of the experimental quantities

$$x_1, x_2, ..., x_m, z_1, z_2, ..., z_n.$$

The relative notion of likelihood ratio is defined in exactly the same way as the relative notion of likelihood. First, if $h$ and $i$ are included in $d$, the likelihood ratio of $h$ to $i$ is the same as their absolute likelihood ratio. Secondly, an equivalence condition. If $d$ implies that $h$ and $i$ are equivalent, the likelihood ratios of $h$ to $j$ and $i$ to $j$ are the same (if either exists at all).

Likelihood ratios might be known even when experimental densities are not. Take a trial $T$ and data $d$ where the result $u$ of $T$ is partly determined by measuring the variable $x$. Suppose $d$ tells nothing about the true value of $u$ at $T$, nor of $x$ at $T$, nor of the value of $x$ at $T$ for given $u$. Then if the experimental densities for hypotheses about $u$ are $g(u)h(x)$, the density for the hypothesis $u = a$ is unknown, but the likelihood ratio given $d$ of $u = a$ to $u = b$ is determined as $g(a)/g(b)$. $h(x)$ cancels out.

Given any data $d$, and any set of propositions $A$, $L$ is a *likelihood function on $A$, given $d$* if for any two members of $A$, $h$ and $i$, $L(h)/L(i)$ is the likelihood ratio of $h$ to $i$ given $d$, or is undefined if that ratio is not uniquely determined by $d$. For brevity we shall treat all constant multiples of $L$ as the same function, and speak of *the* likelihood function, namely a function determined down to a constant factor.

## The principle of irrelevance

If $A$ is a partition of both $d$ and $d \& e$, and the likelihood function on $A$ given $d$ is the same as the likelihood function on $A$ given $d \& e$, then $e$ is called irrelevant to $A$, given $d$.

Finally the *principle of irrelevance* states: *If $e$ is irrelevant to $A$ given $d$, and $h$ is in the logical field based on $A$, then*

$$p(h|d) = p(h|d \& e).$$

It follows that the distribution of support furnished by $d$ over members of the logical field based on $A$, is just the same as the distribution of support furnished by $d \& e$.

*Consistency*

We now have a system based on a sigma-algebra of joint propositions and their components, and employing the usual machinery of set-theory; it includes the Kolmogoroff axioms for chances and the probability axioms for degrees of support; it includes the frequency principle and the principle of irrelevance to link chances and degrees of support. A completely formal analogue of this system can be proved consistent by the logicians' trick of collapsing the full system into the Kolmogoroff axioms. So it has a model in a system currently taken to be consistent. Statisticians may prefer to verify consistency by noting how degrees of support obtained from the principle of irrelevance are dependent on likelihood functions.

Fisher implicitly used a principle of irrelevance which was not consistent with the rest of the axioms. It may be worthwhile to recite a few of the contradictions which arise when his form of the principle is used. The principle I attribute to Fisher is that given initial data $d$ about trials of kind $K$, and any pivotal kind of trial $K'$ defined in terms of $K$ and $d$, any further information $e$ is irrelevant, given $d$, to hypotheses about the outcome of any designated trial of kind $K'$. I call this principle Fisher's because he seems often to have used it. On one occasion he demanded that the pivotal kind of trial $K'$ should be 'natural', but the meaning was never made clear. Anyway, when students have mentioned contradictions in the fiducial argument, they have usually had in mind the manifest contradictions which follow from this principle. The following few paragraphs are a rather technical account of the contradictions, and are not used in the sequel.

The fundamental difficulty with Fisher's principle of irrelevance is that there may be too many pivotal kinds of trial. On Fisher's principle, this leads to different distributions of support for the same data. Here is the most obvious source of contradiction. However, it is known that in the one parameter case pivotal trials based on a single sufficient statistic are essentially unique.† Hence

† D. Brillinger, 'Examples bearing on the definition of fiducial probability with a bibliography', *Annals of Mathematical Statistics*, XXXIII (1962), 1349–55.

fiducial distributions based on a single statistic do not give conflicting results. But this is no solace, for if there is a single sufficient statistic, we may define pairs, triads, and so on which are jointly sufficient, and pivotal kinds of trial based on these will give different distributions of support, if Fisher's method is followed.† So it has been suggested that the argument should be restricted to minimal sufficient statistics. Evidently such a restriction is entirely *ad hoc*. Nor is it enough, for it has been proved that distributions of support obtained by Fisher's method from a single sufficient statistic may still conflict with the other axioms about support.‡

When there are two or more parameters and correspondingly many statistics the matter is notoriously intractable. Contradictions have been discovered in the simplest case of the Normal distribution.§ The Bivariate Normal is a graveyard. A catalogue of corpses has recently been published.‖ But in all these cases, whether of one or more parameters, the trouble can be traced to Fisher's over-generous assumption of irrelevance. Every time a contradiction has appeared, some assumption has been made which is stronger than our principle of irrelevance.

It is interesting to compare the situation in the fiducial argument with the situation in set theory long ago. One of Frege's notable achievements was to derive arithmetic from concepts and postulates of pure logic. His crucial principle required that if a condition can be stated in purely logical terms, then there is a class of things satisfying that condition. Russell refuted this: contrary to expectation not all conditions expressed in logical notation define a class. Frege needed weaker assertions in order to preserve consistency. It is much the same with the fiducial argument. Fisher thought that all data are irrelevant to hypotheses about the outcome of pivotal trials. In fact, as other workers have shown, irrelevance is not so easily come by. Just as we needed a more adequate characterization of class formation, so we need a more adequate

† R. M. Williams, *The use of fiducial distributions with special reference to the Behrens–Fisher problem*. Cambridge Ph.D. dissertation, 1949. Cambridge University Library dissertation no. 1671.

‡ D. V. Lindley, 'Fiducial distributions and Bayes' theorem', *Journal of the Royal Statistical Society*, B, xx (1958), 102–7.

§ A. P. Dempster, 'Further examples of inconsistencies in the fiducial argument', *Annals of Mathematical Statistics*, xxxiv (1963), 884–91.

‖ At the end of Brillinger's article just cited.

characterization of irrelevance. Each genius proferred a false conjecture. The task of future generations is not to confute the genius but to perfect the conjecture. Our principle of irrelevance is some way along that road.

*Pseudo-distributions*

In discussing the finite case it was shown how the fiducial argument and Bayes' formula may combine to give initial distributions of support. Sometimes we can do this for the continuous case, but not always. Take waiting times at a telephone exchange, where, it will be recalled, a waiting time is an interval between one incoming call and the next. Data $d$ shall be the initial data, saying that on some designated trial $T$ there is a non-zero waiting time, and that the cumulative distribution of chances for waiting times is of the form,
$$F(t) = 1 - \exp(-\theta t), \qquad (7)$$
where $F(t)$ denotes the chance of getting a waiting time less than or equal to $t$, and $\theta$ is an unknown constant greater than 0. $\theta$ is a property of the telephone exchange and the community it serves; statistical hypotheses about the set-up are just hypotheses about $\theta$.

The fiducial argument can be shown to apply here; it works even if only a single non-zero waiting time $u$ has been observed. If $e$ states that $u$ was the waiting time at trial $T$, we may obtain a distribution of support furnished by $d \& e$ over hypotheses about the value of $\theta$. This is most easily represented by its density function $f(\theta|d \& e)$,
$$f(\theta|d \& e) = u \exp(-\theta u). \qquad (8)$$
This will be called the posterior distribution of support, since it is the relevant distribution after $u$ has been observed. If there were also an initial distribution of the support furnished by $d$ alone, its density function would be represented by $f(\theta|d)$. It would have to satisfy the continuous version of Bayes' formula,
$$f(\theta|d \& e) = \frac{f(\theta|d) f(e|\theta \& d)}{\int_\theta f(\theta|d) f(e|\theta \& d) \, \mathrm{d}\theta} \qquad (9)$$

The solution is,
$$f(\theta|d) = 1/\theta. \qquad (10)$$
But this can hardly be the density function for an initial distribution of support, for its integral over all possible values of $\theta$ does not

converge. The degree of support for what is certain on $d$ would not be 1, as demanded by our axioms, but no finite number at all. It is at most a pseudo-distribution with no clear interpretation.

What does this mean in our theory of support? (9) holds only if $f(\theta|d)$ is defined within our axiomatic system. It turns out that no distribution of support satisfying our axioms also satisfies (9). Hence $f(\theta|d)$ is not defined in our system of support. There is no initial distribution of support over hypotheses about the value of $\theta$.

The conclusion is certainly plausible, and may even serve as evidence that our system is sound. For the initial data $d$ say only that $\theta$ is positive. $\theta$ has an infinite range, from 0 up. How, on this limited data, to express our alleged indifference between hypotheses about different values of $\theta$? I think no way is uniquely compatible with our axioms. Only when more is learned, only, say, when we learn of a single observed waiting time, can we get significant measures of support for hypotheses about $\theta$. A single observed waiting time gives some vague idea about the location of $\theta$, and we can begin to measure which locations are better supported than others. But before that, there is just no measure of support.

Jeffreys takes the very opposite view. He thinks the pseudo-distribution whose density is $1/\theta$ exactly expresses our ignorance about the location of $\theta$. He has no qualms that the degree of support for what is certain is not 1 but $\infty$. Nor need he: you can measure support on any scale you please. But then he plugs this distribution into formula (9) and arrives at our posterior distributions, where support is measured between 0 and 1. We switch scales in mid-formula. No meaning attaches to this operation. One cannot tell within what set of axioms Jeffreys is working.

From a purely formal point of view, that is, from the point of view of manipulating signs, it does not matter whether you begin with the principle of irrelevance, or take Jeffreys' pseudo-distribution as primary. But if one is concerned with the philosophy of the thing, with the meaning of the operations and a sound foundation for the inferences, it seems impossible to adopt Jeffreys' technique. Indeed, concerned with the logic of the thing, you may even regard the principle of irrelevance as validating a proof that, in some cases, Jeffreys' initial distributions do not exist. Yet the posterior distributions do exist, and coincide with those derived by Jeffreys. Many students may have felt both that

Jeffreys' initial distributions are unfounded, and yet that his posterior distributions are plausible. The fiducial argument, where it applies, exactly answers to this opinion. If the test of a postulate is entailing what is probably true while avoiding what is probably false then, in this regard, the principle of irrelevance fares rather well.

*Completeness*

It is possible to prove, at least for the one-parameter case, that the principle of irrelevance is the strongest general principle which is consistent with the other axioms. I call a principle general if it does not refer to particular characteristics of chance set-ups; in logic, this would be spelled out in terms of the proper names and bound variables occurring in the statement of the principle. Any general principle stronger than our principle of irrelevance inevitably leads to contradiction. Thus the principle of irrelevance completely characterizes the general concept of irrelevance for the one-parameter case. It is to be conjectured that this holds for the many-parameter case, but this has not yet been proved.

This may have consequences for Jeffreys' theory. For it is possible to take *ad hoc* initial distributions or pseudo-distributions which give measures of support in cases where the fiducial argument cannot be applied. These sometimes imply some judgement of irrelevance. But these judgements are in principle not generalizable, for any thorough-going generalization of them would entail a contradiction. So our completeness result may be taken as indicating that where Jeffreys' solutions, in one-parameter cases, go beyond the fiducial argument, they are necessarily *ad hoc*. This does not mean they are wrong, only that there is no extension of them on the same level of generality as the principle of irrelevance.

Now for a sketch of the completeness proof. The one-parameter case is where $d$ states that the cumulative distribution function for trials of kind $K$ is of the form $F(x,\theta) = a$. What is the chance of getting a result less than or equal to $a$ on a trial of kind $K$? It is evidently just $a$. This is independent of the true value of $\theta$. Thus $F(x,\theta)$ is itself a pivotal function.

When does the principle of irrelevance permit use of this pivotal function to determine posterior distributions of support? Let data $e$ state an observed value of $x$. The principle requires that the likelihood function, on the basis of $d\&e$, is the same as the likelihood function, on the basis of $d$. This will happen if and only if the latter function is uniquely determined by $d$, that is, if and only if,

$$\frac{\partial F}{\partial x} = g(F)\, h(x). \tag{11}$$

All solutions of this equation are of the form

$$F(x,\theta) = G(R(x)+S(\theta)). \tag{12}$$

Moreover it can be shown that all pivotal functions give the same posterior distributions as $F$. So (12) states the necessary and sufficient condition that the fiducial argument be applicable in the one parameter case.

But Lindley has proved that (12) represents the necessary condition that the fiducial argument be consistent with the rest of our axioms.† Thus, if the fiducial argument did apply to any more situations than those permitted by the principle of irrelevance, we should have an inconsistency. So in the one-parameter case the principle of irrelevance completely characterizes the general concept of irrelevance.

## Application: theory of errors

Lindley's result suggests an important application of the fiducial argument. When measuring a quantity $\theta$, some measuring device takes readings; these may be conceived as the result of trials on a chance set-up. Suppose that the distribution of chances for different possible readings is some known function of $\theta$, expressed by the cumulative function $F(x-\theta)$. $F$ is a characteristic of the measuring device. Evidently $F(x-\theta)$ is of the form studied in the preceding section, so if it is suitably continuous and differentiable the fiducial argument applies. Hence, given a set of readings, summed up in data $e$, we may state the degree to which $d\&e$ supports any hypothesis about $\theta$. Thus on the basis of our data we have a measure of the reliability of different hypotheses about $\theta$. The fiducial argument thereby solves the fundamental problem of the theory of errors.

† D. V. Lindley, 'Fiducial distributions and Bayes' theorem', *Journal of the Royal Statistical Society*, B, XX (1958), 102–7.

*Application: the Normal family*

It may be worth working through a fiducial argument for the Normal family. This will be done in full, avoiding obvious short-cuts; the present aim is solely to repeat our display of the logic of the argument.

Suppose initial data affirm a Normal distribution of chances on trials of some kind, but that mean and variance are unknown. Intuitively speaking, the result on a single trial on the set-up might give some idea of the mean, but could give no hint as to the dispersion of results, that is, of the variance. Hence, *pace* Jeffreys, there would be no distribution of support over hypotheses about the mean and variance. Support requires more than ignorance. Likewise, two trials yielding exactly the same results might suggest a small overall dispersion, but could give no quantitative estimate of its size. Only when two different results are to hand can we begin the business of curve fitting. Indeed, if the results of different trials were regularly identical, we should not be in the realm of chance but of causal uniformity.

The fiducial argument caters admirably to these miscellaneous hunches. If the initial data do not include the assumption or observation that two different trials have different results, no distribution of support is well defined. Hence we are led to consider trials of kind $K$, and a body of data $d$ which says, first of all, that trials of kind $K$ are independent with Normal distribution and secondly, that two trials, $T_1$ and $T_2$, have different results. Then we are going to take further data $e$, stating the actual results of trials $T_1$ and $T_2$, $a$ and $b$, say, and compute the distribution of support over hypotheses about the mean and variance.

The density function for the Normal distribution is of the form:

$$\frac{1}{\sqrt{(2\pi)}\sigma} \exp\left[-\frac{1}{2}\left(\frac{x-\mu}{\sigma}\right)^2\right] dx. \quad (13)$$

Considering a compound trial consisting of two independent trials, we get a density function for compound trials of the form

$$\frac{1}{2\pi\sigma^2} \exp\left[-\frac{1}{2}\left(\frac{x_1-\mu}{\sigma}\right)^2 - \frac{1}{2}\left(\frac{x_2-\mu}{\sigma}\right)^2\right] dx_1 dx_2. \quad (14)$$

## APPLICATION: THE NORMAL FAMILY

Here $x_1$ represents the result of the first trial in the compound, and $x_2$ of the second. $\mu$ is the unknown mean and $\sigma^2$ the unknown variance. Now for the fiducial argument.

*Step 1.* Construct a pivotal kind of trial and pivotal function. This we do by means of two intermediate functions, $v$ and $t$, defined in terms of $x_1$, $x_2$, $\mu$ and $\sigma$

$$v = \frac{x_1 - x_2}{2\sigma}, \quad t = \frac{x_1 + x_2 - 2\mu}{2\sigma}. \tag{15}$$

These functions define a derivative kind of trial, whose possible results are the possible values of $v$ and $t$. If the two trials of kind $K$ have results $x_1$ and $x_2$, and the true mean and variance are $\mu$ and $\sigma^2$, then the corresponding derivative trial has result $v$ and $t$. Evidently the density (14) may be expressed in terms of $v$ and $t$; direct substitution gives

$$\frac{1}{\pi} \exp[-(t^2 + v^2)] \, dv \, dt. \tag{16}$$

Hence we know the cumulative distribution for $v$ and $t$, namely

$$G(v, t) = \int^t \int^v \frac{1}{\pi} \exp[-(t^2 + v^2)] \, dv \, dt \tag{17}$$

which is entirely independent of the unknown mean and variance. Hence our derivative kind of trial is indeed pivotal, and we have completed step 1.

*Step 2.* Construct a partition of $d$ and $d\&e$ based upon our pivotal kind of trial. The obvious partition to try consists of joint propositions of the form.

$h_{v,t}$: Normal distribution for trials of kind $K$; outcome of trial derivative on $T_1 T_2$ is $v, t$. (18)

This is a partition of both $d$ and of $d\&e$, since truth of either of those implies that some $h_{v,t}$ must be true.

*Step 3.* Discover the likelihood function on our partition, given $d$. Since every $h_{v,t}$ is included in $d$, the likelihood ratio between any two members of the partition, given $d$, will simply be their absolute likelihood ratio, namely the ratio of their experimental densities. The experimental density of any $h_{v,t}$ is as

$$\frac{v^2}{(x_1 - x_2)^2} \exp[-(t^2 + v^2)] \, dx_1 \, dx_2 \tag{19}$$

for all positive and negative $v, t$. $d$ ensures that $x_1 - x_2$ differs from 0. Hence any likelihood function as a function of $v$ and $t$ is simply the second factor

$$v^2 \exp[-(t^2 + v^2)] \qquad (20)$$

multiplied by a constant.

*Step 4.* Discover the likelihood function on our partition, given $d \& e$. Here we observe that no $h_{v,t}$ is included in $d \& e$—every $h_{v,t}$ admits of some possibilities not admitted by $d \& e$. So we cannot assume that the likelihood ratio between two $h_{v,t}$ is given by their absolute likelihood ratio. But data $e$ says that the result of trial $T_1$ is $a$, and of trial $T_2$, $b$. Hence given $d \& e$, each $h_{v,t}$ is equivalent to a joint proposition,

$h^*_{v,t}$: Distribution is Normal on trials of kind $K$, with

$$\sigma = \frac{a-b}{2v} \quad \text{and} \quad \mu = \frac{a+b-2\sigma t}{2};$$

the result of trial $T_1 T_2$ is $a, b$.

Each $h^*_{v,t}$ is included in $d \& e$; so their likelihood ratios given $d \& e$ will simply be their absolute likelihood ratios, and it is easy to check that the likelihood function over the partition consisting of the $h^*_{v,t}$—and hence of the equivalent partition consisting of the $h_{v,t}$—is as given in (20).

*Step 5.* Check that the likelihood functions given $d$ and given $d \& e$ are the same, so that $e$ is irrelevant to the partition, given $d$. In virtue of step 2, we know the distribution of support over hypotheses about $v$ and $t$, given $d$. Now we know it given $d \& e$. And since every hypothesis about the mean and variance is, given $d \& e$, equivalent to an hypothesis about $v \& t$, we have thus a distribution of support over hypotheses about the mean and variance.

The distribution which results is of no especial interest to this abstract study; suffice to say that it exactly coincides with the one Fisher urged all along. But Fisher took steps 3, 4, and 5 for granted. It is hardly surprising that if these steps are omitted, you can get other pivotal functions which will give different, contradictory distributions of support. It is no use protesting, as Fisher did, that those functions are not 'natural'. What you have to do is give a theory prescribing what is natural. Such is our theory. It insists on the intermediate steps, and so prevents contradiction.

## Application: confidence intervals

Neyman's theory of confidence intervals has long been esteemed by statisticians, both for the elegance of its mathematics and the lucid quality of Neyman's original exposition.† It is in a way the inverse of the Neyman–Pearson theory of testing, and it introduces no new philosophical questions. But it may be helpful to examine its relation to the fiducial argument. This is especially necessary since Fisher's original expression of the fiducial argument is couched in terms more suitable to Neyman's theory than his own.

First consider a special kind of function, to be called a *Neyman function*. Let there be a chance set-up $X$, trials of kind $K$, some initial data $d$ stating that the distribution of chances on trials of kind $K$ is a member of the class $\Delta$. A Neyman function will be, first of all, a function from results of trials of kind $K$ to subclasses of $\Delta$; it is a function of the form $f(E)$, where for all possible results $E$, $f(E) \subseteq \Delta$.

In terms of such a function we define another. Say $g(E)$ has the value 'inclusion' if $f(E)$ includes the true distribution, and has the value 'exclusion' otherwise. $f$ is called a Neyman function if and only if $g$ is pivotal, namely if the chance of inclusion is independent of the true distribution. When $f$ is a Neyman function, the chance of inclusion is called the *confidence coefficient*.

With a Neyman function of confidence coefficient 0·95, the chance of getting some result $E$, such that $f(E)$ includes the true distribution, must be 0·95. By the frequency principle, 0·95 is a measure of the degree to which the initial $d$ supports the hypothesis, that inclusion will occur. You could say, in the light of $d$, and before making a trial on the set-up, that you can be 95% confident that inclusion will occur.

In Neyman's theory, if you make a trial on the set-up and get result $E$ for which $f(E) = \Gamma$, then $\Gamma$ is an appropriate 95% *confidence interval*. If no further significance is attached to confidence intervals no harm is done: they are simply classes of distributions computed according to certain principles. But it has

† J. Neyman, 'Outline of a theory of statistical estimation based on the classical theory of probability', *Philosophical Transactions of the Royal Society*, A, CCXXXVI (1937), 333–80.

been thought that you can be 95% confident that $\Gamma$ does include the true distribution. Neyman may never have said so, but other writers seem to have computed confidence intervals just because they favoured this interpretation. It is easy to show the interpretation wrong by exhibiting cases in which you can be certain on some data that a 95% confidence interval computed for that data does *not* include the true distribution. The very cases used in discussing the Neyman–Pearson theory of testing could be reproduced here, but such repetition would be tedious.

A 95% confidence interval is really worthy of 95% confidence on the data—or, at least, is an interval supported to degree 0·95—only where the principle of irrelevance can be used. Thus confidence intervals can be used for after-trial evaluation only if criteria of irrelevance are satisfied. When they are satisfied use of confidence intervals coincides with use of the fiducial argument. When the conditions are not satisfied, it is not clear that confidence intervals are of much value to statistical inference. Neyman defended them from the point of view of what he called inductive behaviour. The issues raised here have already been discussed in connexion with Neyman–Pearson tests.

CHAPTER X

# ESTIMATION

Estimation theory is the most unsatisfactory branch of every school on the foundations of statistics. This is partly due to the unfinished state of our science, but there are general reasons for expecting the unhappy condition to continue. The very concept of estimation is ill adapted to statistics.

The theory of statistical estimation includes estimating particular characteristics of distributions and also covers guessing which possible distribution is the true one. It combines *point estimation*, in which the estimate of a magnitude is a particular point or value, and *interval estimation*, in which the estimate is a range of values which, it is hoped, will include the true one. Many of the questions about interval estimation are implicitly treated in the preceding chapter. A suitable interval for an interval estimate is one such that the hypothesis, that the interval includes the true values, is itself well supported. So problems about interval estimation fall under questions about measuring support for composite hypotheses. Thus the best known theories on interval estimation have been discussed already. The present chapter will treat some general questions about estimation, but will chiefly be devoted to defining one problem about point estimation. It is not contended that it is the only problem, but it does seem an important one, and one which has been little studied. When the problem has been set, the succeeding chapter will discuss how it might be solved. But neither chapter aims at a comprehensive survey of estimation theory; both aim only at clarifying one central problem about estimation.

*Guesses, estimates, and estimators*

An estimate, or at least a point estimate, is a more or less soundly based guess at the true value of a magnitude. The word 'guess' has a wider application than 'estimate', but we shall not rely on the verbal idiosyncracies of either. It must be noted, however, that in the very nature of an estimate, an estimate is supposed to be close to the true value of what it estimates. Since it is an estimate of a magnitude, the magnitude is usually measured along some scale, and this scale can be expected to provide some idea of how close the estimate is. Estimates may be high or low; there are over- and underestimates; but guesses on the other hand are not usually said to be high or low; we have no English word 'overguess'. This is because the concept of guessing is not geared to scales and magnitudes. The notion of the closeness of a guess cannot be expected to apply as widely as the closeness of an estimate.

An estimate can be called good for two different reasons. After the true value of what is being estimated is known, an estimate might be called good if it is in fact close to the true value. But whether or not the true value is known, an estimate based on certain data may be called good if the data give good reason to believe the estimate will be close to the true value. The former kind of excellence is laudable, but the whole point of estimation is that we estimate before true values are known. Hence we must choose estimates not because they are certainly close to the true value, but because there is good reason to believe they will be close.

To avoid confusion, an estimate based on data, such that the data give good reason to believe the estimate is close to the true value, will be called a well-supported estimate. Similar locutions will also be used. Indeed, since our inquiry is about choosing just such estimates, when the meaning is clear it is occasionally convenient simply to call such estimates good.

Study of estimates requires attention not only to particular estimates, but also to systems for computing or deriving estimates of various magnitudes from various bodies of data. Such a system will be called an *estimator*. Essentially an estimator is an estimate function which maps data on to estimates appropriate to that data. A *good estimator* will be an estimator which provides a well-supported estimate, for any body of data with which it can deal.

Thus by a good estimator we shall not mean one which with certainty gives estimates each of which is close to the true value. If there were, in statistical problems, such estimators they might be called close estimators: but there are not any. In all interesting problems any estimator might be wildly wrong.

Many statisticians take quite a different view on estimators. We call an estimator good if it it gives well-supported estimates, for every body of data it can handle. And a well-supported estimate for given data is one which the data give good reason to believe will be close to the true value. But many statisticians, partly because they lack any solid logic of support, believe that statistical data can never give reason to believe any particular estimate is close to the true value. All that can be done, they say, is to choose estimators which on the average, or very often, give estimates which are in fact close to the true value of what is under estimate. It is possible, of course, that we too shall reach this conclusion. But we should at least begin inquiring after well-supported estimates, rather than estimators which are close on the average.

There are of course practical problems where one wants an estimator which, on the average, gives answers close to the true value. But there are other problems in which a well-supported estimate is desired. A textile manufacturer may want some general method for estimating a property of batches of linen thread, and he may be content with a method which is, on the average, close to the true value. But another man, requiring a particular batch of thread for a particular and unique purpose, does not necessarily want a method which is close on the average; he wants a good estimate, that is, an estimate which is well supported by the data available about the thread. Of course the two estimates might coincide, but we shall show that in general they need not. Even if estimator $T$ is closer on the average than estimator $S$, it can happen that in particular situations, $T$'s estimate is less well supported than $S$'s. So we must keep the two conceptions distinct.

The distinction between average properties, and being what I call well-supported, is similar to one we have already made in the theory of testing. It resembles the contrast between before-trial betting and after-trial evaluation. Before making a trial, it may be wise to favour an estimator which is close on the average: it is a good before-trial bet. But a further question concerns whether this estimator is suitable for after-trial evaluation: for making an estimate which is good in the light of data discovered by a particular trial.

*Belief and action*

An earlier part of the essay gave reasons for not embarking on decision theory as the first step in elucidating chances. Decision theory is the theory of what action to take when the consequences of various possible alternative actions are known. It has seemed to some writers that making an estimate is performing an action and that the consequences of various estimates can be evaluated. So the problem of choosing the best estimate is essentially one of deciding which action to take, and hence falls under decision theory. This approach has sometimes been criticized on the ground that often in science the consequences of estimates are beyond evaluation. But this objection grants far too much. My curter reasons for not studying estimation under the head of decision theory are (*a*) *estimates do not have consequences* and (*b*) *estimating is not acting*.

To begin the explanation, I should like to ignore estimation altogether, for as future sections will show, guessing and estimation have ambiguities all of their own. First it is useful to distinguish two families of ideas. Belief is in one family, action in the other. Estimation, I should say, is in the same family as belief, but it will be left aside for a few paragraphs.

*Beliefs do not have consequences in the same way in which actions do.* This is a logical cum grammatical point about the concepts of belief and action. Indeed we say that a man did something in consequence of his having certain beliefs, or because he believed he was alone. But I think there is pretty plainly a crucial difference between the way in which his opening the safe is a consequence of his believing he was unobserved, and the way in which the safe's opening is a consequence of his dialling the right numbers on the combination. It might be expressed thus: simply having the belief, and doing nothing further, has in general no consequences, while simply performing the action, and doing nothing further, does have consequences.

There is a related difference: *It is reasonable for a man to perform a deliberate action only if he takes account of the consequences of performing that action, but it is in general reasonable for a man to have an explicit belief without his taking account of the 'consequences' of having that belief.* This fact is probably just a reflexion of the different way in which beliefs and actions have consequences. It is reasonable to have an explicit belief without taking account of its consequences just because it does not have consequences in the way in which an action has consequences.

It must be noted that neither of the two italicized assertions denies that the correct way to analyse the concept of belief may be in terms of action and dispositions to act in various ways. Nor does it deny that the sole way of knowing what a man believes may be through a study of his actions. It has perhaps been thought that because belief must be analysed in terms of dispositions to act, beliefs have consequences in just the way actions do. But this is false; beliefs may be analysed in terms of dispositions to act, but beliefs are not actions. Finally, I do not deny that a man may, with some semblance of reason, come to believe something because he thinks his believing it will have good results. We are invited to do this very thing in the game-theoretic argument for belief of God's existence which is known as Pascal's wager. Believe, Pascal advises, for if God does not exist you will lose little through the false belief that he does exist; but if God does exist and you do not believe, you court eternal damnation. Here we are pretty well in the domain of belief in something, rather than belief that something is so; we are near to where 'acts of belief' and the evangelist's 'decision for Christ' do make sense. They have only a little to do with believing *that* something is the case.

## A general consideration

It follows from the italicized assertions of the last section that reasonable decision to perform an action must involve considerations different from the formation of reasonable beliefs. I wish to urge that exactly the same is true of estimation, that an estimate may be shown good without considering any consequences of anything. But we must pay special attention to features of guessing and estimation before accepting this thesis.

A general consideration may prove helpful before proceeding. There are often pretty good reasons for our having the words, and the distinctions marked by words, which we do in fact have. Indeed with the progress of rational investigation these good reasons may be superseded, and the distinctions abolished, but to know whether this should happen in a particular case, it is good to have in mind the general reasons behind the existing distinctions. In the case of belief and action the matter is as follows. Men have been able to investigate the consequences of actions rather well; they have a pretty good understanding of physical laws. They have also made rapid progress in understanding which beliefs are reasonable, and which not. But there has been virtually no progress in understanding the relation between belief and action; here all is chaos and conceptual confusion.

Hence my desire to treat estimation theory separate from decision theory is not only an instance of a fearful conservativism: it stems from a belief that rigour cannot precede clarity. There are plenty of clearly understood sanctions in the formation of belief, and plenty in questions of how to act. But there are none in between. Until some sanctions exist in between there is not likely to be any precise and rigorous theory spanning the two. Nor, for a study of estimation, is one needed.

*Carnap's rule*

Various facts incline various people to believe estimates have consequences. The most important, and the most relevant, is stated in the following section. First another influential consideration must be recalled: the idea that to make an estimate is to decide to act as if the estimate were a true, or almost true, statement of the facts.

This wrong idea is implicit in many statistical writings. Perhaps it stems from accepting a rule like that expressed by Carnap: 'Suppose that your decision depends upon a certain magnitude $u$ unknown to you, in the sense that, *if* you knew $u$, then this would determine your decision.... Then calculate the estimate $u'$ of $u$ with respect to the available evidence $e$ and act in certain respects as though you knew with certainty that the value of $u$ were either equal to $u'$...or near to $u'$.'† This may be all right if the 'certain respects' are sufficiently feeble. But Carnap later sums up the rule with what he seems to have had in mind all along, 'Act as if you knew that the unknown value of the magnitude in question were equal or near to the estimate'.‡

He retracts this rule in favour of another, but it is important to

† R. Carnap, *Logical Foundations of Probability* (Chicago, 1950), p. 257.
‡ *Ibid.* p. 532.

see that the rule, in its brief form, is absurd. The whole point of estimates is that they are estimates, not certain truths or certain approximations to truth. I estimate that I need 400 board feet of lumber to complete a job, and then order 350 because I do not want a surplus, or perhaps order 450 because I do not want to make any subsequent orders. At any rate I do not act as if I know 400 to be my real needs. Nor need I act as if it is near my true needs; there is no reason why I should not, (*a*) count any estimate as far out if it misses the true need by more than 10 board feet, and (*b*) order 350 or 450 just because I fear that my estimate may be very far out. 400 is my estimate, but I do not act as if it is near my true needs. Some distinguished schools of engineering have taught their pupils to apply a correction factor of *eight times*: first estimate, to the best of human abilities, the structural strains on the projected design, and then multiply by a factor of eight: act as if your estimate were eight times wrong. Of course you could stretch the interpretation of Carnap's rule, saying that 350 is in some sense near my true needs. Then the rule is empty. You can always find an occasion to say Cambridge is near London, the earth is near the moon, and the sun is near $\alpha$-Centauri.

Hence it is not a good universal rule to act as if I knew an estimate were near the true value of what it estimates. This should make plain how absurd is the much stronger claim, that to make an estimate is to decide to act as if the estimate were a true, or almost true, statement of the facts. The stronger claim would be false even if Carnap's rule were excellent; it is obviously false since the rule is itself a bad one.

There is a closely connected confusion between estimates and acts based on estimates. To take but one example, Savage considers a person who is to 'estimate the amount $\lambda$ of shelving for books, priced at $1.00 a foot, to be ordered for some purpose'.† He proceeds to describe, not a method for estimating, but one for deciding how much to order. Deciding how much to order is not the same as estimating the amount needed. For as already remarked, I can perfectly well estimate 400 feet, and then order 350 so as not to incur wastage, or 450 so as not to need two orders. Savage writes as if 350 or 450 would be my estimates, but this is wrong. They are my orders.

† L. J. Savage, *The Foundations of Statistics* (New York, 1954), p. 232.

Unfortunately these confusions have a deeper source than may at first be noticed. In the very idea of estimation and of guessing there is an ambiguity which leads to a conflict between the idea of a guess and the act of guessing, between the estimate and the act of estimation. It needs a detailed section to explain the matter.

*Guesses and acts of guessing*

The word 'estimating' may refer to two slightly different things. So may 'guessing'. If I say I guess so and so, I am both performing an act and giving expression to something like an opinion or a belief. Indeed the act is the less important to determining whether someone really did guess: we say, 'he only said he guessed $A$, but it was not his real guess', or, 'he said he guessed $A$, but did not really guess at all'. Yet in some situations a man's saying he guesses something may make that his guess.

If a foolish person asks me to guess, from blurry newspaper photos, which of two women is the more beautiful, I may discreetly guess it is his favourite even if that is not my real opinion; if he asks me to estimate the length of his new and extravagant yacht, I may tactfully estimate it to be longer than I really suppose it to be. Although in my private opinion, the length is about 30 metres, I say I guess 35. Perhaps I only pretend to guess 35, but it is not obviously wrong, or even necessarily misleading, to say that in the company of the financier, I estimate that his yacht is 35 metres long. The possibility of using the words 'guess' and 'estimate' in this way presents a severe difficulty. For there could now be two distinct grounds for judging whether an estimate is 'good' or not. As we have said, a guess based on certain information is a good one if the information gives reason to suppose the guess is right. Guesses aim at the truth; likewise an estimate of a magnitude aims at being near the actual magnitude. But we may also judge a person's *act* of guessing, or his act of estimation. Even if a guess is certainly near the truth, it may be very unwise to express it; it may be foolish, as one might say ambiguously, to make the guess. This is most evident when making the guess known may have noxious consequences.

Statistical examples show this double standard is not restricted to social discretion. Consider a case drawn from a preceding chapter. A man must guess on the constitution of each of a long sequence of unrelated urns. Each urn may have either (*a*) 99 green balls, 1 red; (*b*) 2 green, 98 red; (*c*) 1 green, 99 red. It is known that the chance of drawing a ball of any colour from an urn is in proportion to the number of balls of that colour in the urn. But the precise order and arrangement of the urns is unknown. The first might be (*a*), the next (*c*), or vice versa; all might be (*b*). Our man's job is to draw a ball from an urn, guess its constitution and proceed to the next. The urns come from different sources, so the constitution of the urns he examines is no key to the ones not yet examined.

We remarked that under these circumstances, a man might think it foolish to guess (*a*) when green is drawn, and (*c*) when red is drawn, for he might be wrong every time: there is nothing in the data to say the urns are not all (*b*). A mixed strategy seems preferable. Aiming at being right as often as possible, our man decides to minimize his maximum possible chance of error, and computes that when green is drawn, he should flip an almost fair coin to decide whether to guess (*a*) or (*b*); likewise when red is drawn, he flips the coin to decide whether to guess (*c*) or (*b*). Let us not discuss whether this is the best policy he can follow, but grant only that it is not a foolish one.

It seems quite correct to describe the case as I have done: the man flips a coin, and decides to *guess* accordingly. He adopts a policy for guessing; we imagine him receiving a prize every time he guesses right, and forfeiting a small stake for every error. Yet in speaking in this way we are essentially considering his acts.

For although the man acts in this way, he might not at all believe that when the coin tells him to guess (*b*), the urn really is (*b*). After all, if green is drawn and the urn is (*b*), a pretty rare event has happened; if the urn is (*a*), only the most commonplace thing has occurred. He may be pretty confident the urn is really (*a*)—he may indeed, if we may say so, guess the urn is (*a*). But he is not entirely confident the urn is not (*b*) or (*c*). So he follows the minimax policy and says he guesses (*b*).

This is not a case of deceit, of pretending to guess (*b*) and really guessing (*a*), or even of first guessing (*a*) and then guessing (*b*). It is hard to put the situation briefly. Roughly, the man guesses the urn is (*a*), but because he is not quite confident, he follows a policy which instructs him to make the guess that the urn is (*b*).

His making the guess (*b*) is consistent with its being correct to say he still guesses the urn is (*a*).

Some readers will want to say that (*a*) is the man's real guess about the urn while (*b*) is only what he says. Others will insist that (*b*) is the real guess, while (*a*) is only a figment of the man's fancy. Best to abandon merely verbal issues, and adopt a new terminology.

In most of what follows I shall call something a *guess about A* only if it aims at the truth about $A$ itself. Occasionally, for explicitness, this will be called a belief-guess. In this way of speaking if a man as it were makes a bet about $A$, not aiming at the truth about $A$, but aiming at being right as often as possible in 100 specified similar guesses about things like $A$, then he is not necessarily guessing about $A$. I shall say instead that he performs the *act of guessing* about $A$. It is evident that an act of guessing should be made with regard to the consequences of making the act. But a guess about $A$, as I so use the term, need not be related to consequences. Thus in the dilemma of the urns, a man might have (*a*) as his guess about the urn, and still make the act of guessing (*b*). There is no question of deceit here, as there is when I lie to absurd yachtsmen. A man might say, 'My guess about the urn is that it is (*a*), but since there is a pay-off over the whole sequence of guesses I make, then formally, in this game, I can see it is wiser to make the guess (*b*)'.

Likewise I shall distinguish an estimate of a magnitude from an act of estimating a magnitude; sometimes for greater clarity I shall call the former a belief-estimate. Notice that neither belief-guesses nor belief-estimates have been explained as something private and incommunicable. The condition on a belief-guess is that it should aim at the truth about what is guessed on, rather than have some wider aims, like pleasing a friend or winning more credits in the long run. A guess is a belief-guess if it aims at the truth regardless of the consequences. I wish to attach no moral overtones to this; I am afraid it sounds too primly commendable to aim at truth regardless of consequences; one should never forget that sometimes it may be stupid to act solely with the aim of stating the truth. But regardless of ethics or expediency we may still have the conceptual distinction between a belief-guess and an act of guessing.

## Scaling

We cannot altogether ignore the purpose to which an estimate will be put, not even a belief-estimate. For the purpose seems to determine the scale upon which a magnitude will be measured, and what appears close on one scale may look distant on another. Hence what seems to be a good estimate on one scale may seem to be a bad one on another.

Carnap calls the scaling problem the paradox of estimation.† A form of the difficulty can be put as follows. Suppose we are estimating how far a point on a great circle is from the North Pole. The distance may be expressed in many ways. There are miles. There is angular distance measured between $-180°$ and $+180°$. There is positive angular distance measured from $0°$ to $360°$. Suppose the distance is estimated at $+5°$, while it is in fact $-5°$. Then on one scale the error is a mere $10°$; on the other, it appears as $350°$. However, this example may seem a sort of trick. So consider cases in which a magnitude is measured first on a scale of miles, then on a scale of miles squared, then on a scale of the sine of the angular difference. For some distances what is a relatively small error on one scale is relatively large on another.

The difficulty can be put another way. It seems plausible to suppose that if $\lambda$ is an estimate of $\theta$, while $\mu$ is an estimate of $\theta^2$, and if these are respectively the best estimates to be based on current data, then $\mu$ should equal $\lambda^2$. But since what is close on one scale may be relatively distant on another, the best estimate on one scale need not coincide with the best estimate on another.

One might answer this dilemma by saying choice of the scale depends on the purpose at hand, and that every study of estimation must thus explicitly consider purposes. Carnap believes that in practical matters scales should be reduced to a scale of value; we should measure the error of an estimate in terms of the losses in money or of goods, which are incurred by making the estimate. So no paradox. This is perhaps a fitting recipe for acts of estimation, but to apply it to estimates is to confuse estimates with acts of estimation. After all, even the most practical man can accurately describe two estimates as equally close to the true value of something, yet add that one is a disastrously costly underestimate (the

† R. Carnap, *Logical Foundations of Probability*, pp. 530 ff.

bridge collapsed) while the other is a happily economical overestimate (the bridge withstood a gale of unforeseen proportions).

Attention to ordinary, unstatistical and unsophisticated estimates probably provides the correct answer. Asked to estimate the cost of a house or the length of a ship or the temperature of the day or the melting point of a new alloy, one knows the scale desired, or at least a family of scales which are mere linear transformations of each other (as Fahrenheit is a linear transform of Centigrade, metres of feet, pounds of francs). Indeed if a scale is not stated, and one is asked to estimate the damage caused by, say, a natural catastrophe, one is hard put to answer, and will, perhaps, estimate along several different scales, or else say the damage cannot be estimated. But if one be asked to estimate the loss of life, or the number of buildings destroyed, or the value of property damaged, then estimates can be made along one scale or another. This suggests that the request for an estimate should carry with it, explicitly or implicitly, a request for the scale on which the magnitude under estimate is measured.

This implies not that the theory of estimation must include a theory of which scale to choose, but the theory should take for granted scales which are chosen on grounds independent of the theory. For one purpose one scale may be fitting, for another, another. In what follows we shall consider only estimates of magnitudes when the scale for measuring the magnitude is assumed; we shall take this scale as a measure of closeness.

*Invariance*

According to this analysis, the best estimate of $\theta$ does not automatically imply an estimate of $\theta^2$, so, on this analysis, there is no need to demand that all estimators be invariant under various kinds of transformation. However, although it is not universally necessary, in fact our criteria for goodness of estimators will be invariant under large families of standard transformations; this should mollify many students who have assumed that estimation theory must be independent of particular scales and measures.

One reason for the traditional emphasis on invariance is an attitude to estimation very different from that expressed here. I have been supposing that estimates aim at being close to the true value of some magnitude. This is, I believe, a correct account derived from what the English word 'estimate' means. But in Fisher's opinion, an estimate aims at being an accurate and extremely brief summary of the data bearing on the true value of some magnitude. Closeness to the true magnitude seems to be conceived of as a kind of incidental feature of estimates.

Thus from Fisher's point of view, he is quite correct when he writes, 'if an unknown parameter $\theta$ is being estimated, any one-valued function of $\theta$ is necessarily being estimated by the same operation. The criteria used in the theory must, for this reason, be invariant for all such functional transformations of the parameters'.† He is right, for by an estimate he means a summary of the data.

It is often a trifle hard to see why what Fisher calls estimates should in any way resemble what are normally called estimates. And in fact his estimates based on small samples consist of a set of different numbers, only one of which resembles what we call an estimate, and the others of which are what he calls 'ancillary statistics', namely supplementary summaries of information.

Thus from Fisher's point of view, estimates should be invariant under functional transformations; under our analysis of the notion of an estimate, this is not true. However, in statistics the fundamental criteria for choice of what in this book are called estimates, are in fact invariant under a large family of transformations. In particular they are invariant under linear transformations.

† R. A. Fisher, *Statistical Methods and Scientific Inference*, p. 140.

CHAPTER XI

# POINT ESTIMATION

Given some data our problem is to discover the best point estimate of a magnitude. For this purpose a good estimate will be what I have called a *well-supported* one: an estimate which the data give good reason to suppose is close to the true value. We assume that a scale of closeness is implied in the statement of the problem. We shall be concerned with the excellence of individual estimates, and of estimators making estimates for particular problems, rather than with comparing the average properties of estimators. We seek estimators which give well-supported estimates.

Only rather limited and feeble solutions are offered here. Perhaps there is no general answer to fundamental questions about estimation. Maybe estimation, as I have understood it in the preceding chapter, is only barely relevant to statistical inference. One ought, perhaps, to search for other, related, notions which are more fruitful. But first we must try to do our best with well-supported estimates of statistical parameters.

## *An historical note*

A little history will hint at the flavour of chaos in estimation theory. Unfortunately few students have distinguished the property of being a well-supported estimate, from the property of being an estimate made by an estimator which is in some way close on the average. So the traditional estimators I am about to describe were seldom aimed at one or the other of these two properties, but if anything, at both at once.

Estimation in statistics is relatively recent; it followed two centuries of estimation in other fields. Most work did, however, make use of the doctrine of chances. The usual problem concerned measurement. Some quantity, such as the distance of a star or the melting point of a chemical, is measured several times; there are small discrepancies; now use the measurements to estimate the true value of the physical quantity. Another problem is curve

## AN HISTORICAL NOTE

fitting; try to fit a smooth curve to a number of different points all based on observations. This is both a problem of guessing the true curve, and of estimating the characteristics of that curve.

There were forerunners, but Laplace and Gauss are the great pioneers of what came to be called the theory of errors. Despite their many theoretical insights, the whole science remained pretty *ad hoc* until the 1920's. Many issues had to be settled which now are taken for granted. There were long disputes on how error ought to be measured. It is instructive to quote Gauss on this point. He decided that the square of the distance of an estimate from the true value is a fitting indication of the error committed.

> If you object that this is arbitrary, we readily agree. The question with which we are concerned is vague in its very nature; it can only be made precise by pretty arbitrary principles. Determining a magnitude by observation can justly be compared to a game in which there is a danger of loss but no hope of gain.... But if we do so, to what can we compare an error which has actually been made? That question is not clear, and its answer depends in part on convention. Evidently the loss in the game can't be compared directly to the error which has been committed, for then a positive error would represent a loss, and a negative error a gain. The magnitude of the loss must on the contrary be evaluated by a function of the error whose value is always positive. Among the infinite number of functions satisfying these conditions, it seems natural to choose the simplest, which is, beyond contradiction, the square of the error.

He continues by casting aspersions upon Laplace's idea, that the absolute value of the error is more fitting that its square.†

Once Gauss' view was accepted, *the method of least squares* became the standard, especially in curve fitting. Suppose a curve of some known family is to be chosen as the best representation of some function; suppose a number of points on the curve have been measured approximately. The method of least squares instructs us to choose that curve which minimizes the sum of the squares of the distances between the curve, and the measured points. Some of Gauss' ideas by way of justifying this approach underlie the modern rationale of estimation.

Another famous method in curve fitting is due to Karl Pearson; it is called *the method of moments*. The mean value of a function is,

---

† K. F. Gauss, 'Theoria combinationis observationum erroribus minimus obnoxiae, pars prior' (1820); printed in the *Werke* (Gottingen, 1880), IV, pp. 6–7.

speaking loosely, its average value; the first moment about the mean is the 'average' of the distances between the value of the function and its mean; the $n$th moment is the 'average' of the $n$th power of these distances. This is an analogy from mechanical science. Likewise, given an observed set of points purportedly from a curve, their mean is their average, their first moment the average of their distances from the mean, and their $n$th moment the average of the $n$th power of their distances from the mean. In Pearson's method of curve fitting, suppose the true curve must come from some given family; suppose $k$ parameters suffice uniquely to specify any member of the family. Then compute the first $k$ moments of the data, and choose the curve which has just these values for its own first $k$ moments.

Evidently such a treatment is overwhelmingly *ad hoc*. If it is a desirable technique, this awaits proof. Note that Pearson's is not only a method of curve fitting; it directly estimates parameters of a curve. But there was no general theory for comparing different estimators until 1925, when Fisher published his great 'Theory of statistical estimation', a culmination to a sequence of papers which began in 1912.[†]

The best known product of Fisher's research is the *method of maximum likelihood*. This method, first used by Daniel Bernouilli, and also, it seems, by Gauss, gained new stature from Fisher's work. Choose, according to this method, that estimate of the parameter which has, on the data, the greatest likelihood—in our terms, choose as estimate the parameter of the best supported statistical hypothesis. However, Fisher's great contribution to estimation theory is not this method, but conceptual machinery for testing the method and comparing its excellence to other techniques.

Even some of the latest books will state the method of maximum likelihood as if it were self-evidently correct. It is worth noting that Fisher never postulated his method as something obviously right. In 1912 he may have thought it near to self-evident and beyond dispute, but ever after he realized it was not, and endeavoured to prove it had desirable properties. He seems never to

† R. A. Fisher, 'On an absolute criterion for fitting frequency curves', *Messenger of Mathematics*, XLI (1912), 155–60; 'Theory of statistical estimation', *Proceedings of the Cambridge Philosophical Society*, XXII (1925), 700–25.

have claimed that maximum likelihood estimators were self-evidently the best—only that they were demonstrably better than many others.

*Comparing estimates*

One fundamental difficulty about estimates can be perceived without any technical details. It arises no matter how we construe terms like 'support' or 'being good reason for'. It occurs as much in connexion with average properties of estimators as with the notion of being a well-supported estimate. For although I have called an estimate well supported, or simply good, if there is reason to believe it is close to the true value of what is under estimate, this account does not, in general, imply when one estimate is better than another. Nor is there a clear way of defining betterness in terms of this account, or of any other plausible account.

Suppose magnitude $\theta$ is under estimate. It might seem that $\lambda_1$ is a better estimate than $\lambda_2$ if the data give good reason to believe that $\lambda_1$ is closer to $\theta$ than $\lambda_2$. But not necessarily. For the data might give good reason to believe not only this, but also that $\lambda_2$ is never far from the true value of $\theta$, while $\lambda_1$, if not actually closer than $\lambda_2$, is wildly off the mark. In such an event, which is the better estimate? The question is not well defined.

Or to put the matter in more quantitative terms, let $\epsilon_1$ and $\epsilon_2$ be two small errors of different sizes. Suppose the data give more reason to believe that $\lambda_1$ is within $\epsilon_1$ of $\theta$ than that $\lambda_2$ is, yet also give more reason to believe that $\lambda_2$ is within $\epsilon_2$ of $\theta$ than that $\lambda_1$ is. Which estimate is better? There is no general answer. In particular practical problems there may be some way of choosing between these two estimates in terms of usefulness, but that is all.

This is a universal difficulty which arises directly from that aspect of an estimate, that it should be 'close' to the true value of what is under estimate. It has nothing specially to do with how to analyse expressions like 'the data give good reason to believe'. We shall soon see how the problem recurs even for average properties of estimators. Yet something can be salvaged. We can define a relation between estimates in virtue of which one estimate can be called *uniformly better* than another. This provides for a class of admissible estimates: an estimate is admissible if no rival estimate is uniformly better than it. But I am afraid that the class of admissible estimates is very broad. In most problems there will be a whole host of admissible estimates.

*Uniformly better*

Imagine that $\lambda_1$ and $\lambda_2$ are estimates of $\theta$, and that for *any* possible error $\epsilon$, from o on up, the data give at least as much reason to believe that $\lambda_1$ is within $\epsilon$ of $\theta$, as that $\lambda_2$ is within $\epsilon$ of $\theta$. Then, no matter how one understands being 'close' to the true value, there is no reason to believe $\lambda_2$ is closer to $\theta$ than $\lambda_1$ is. $\lambda_1$ is at least as good an estimate as $\lambda_2$. Let us explicate this idea not in terms of reason to believe, but in terms of our logic of support.

Consider some region around $\theta$, say the interval between $\theta - \epsilon_2$ and $\theta + \epsilon_1$ where $\epsilon_1$ and $-\epsilon_2$ may be thought of as positive and negative errors respectively. Then in order that $\lambda_1$ should be at least as good as $\lambda_2$ we require that for every such region defined in terms of errors $\epsilon_1$ and $\epsilon_2$ the hypothesis that $\lambda_1$ is in that region is at least as well supported as the hypothesis that $\lambda_2$ is.

So our first definition is as follows. If $\lambda_1$ and $\lambda_2$ are estimates of $\theta$ based on data $d$ then in the light of $d$ $\lambda_1$ is *at least as good as* $\lambda_2$ if for every pair of positive errors $\epsilon_1$ and $\epsilon_2$, $d$ supports the hypothesis,

$$-\epsilon_1 \leqslant \theta - \lambda_1 \leqslant \epsilon_2$$

at least as well as the hypothesis,

$$-\epsilon_1 \leqslant \theta - \lambda_2 \leqslant \epsilon_2.$$

Moreover, if, in the light of $d$, $\lambda_1$ is at least as good as $\lambda_2$ but $\lambda_2$ is not at least as good as $\lambda_1$, then $\lambda_1$ is *uniformly better* than $\lambda_2$.

When we have quantitative measures of support, represented by '$p(\ |\ )$', the definition may be abbreviated. $\lambda_1$ is uniformly better than $\lambda_2$ in the light of $d$ if for every pair of positive errors $\epsilon_1$ and $\epsilon_2$,

$$p(-\epsilon_1 \leqslant \theta - \lambda_1 \leqslant \epsilon_2 | d) \geqslant p(-\epsilon_1 \leqslant \theta - \lambda_2 \leqslant \epsilon_2 | d)$$

with strict inequality between the main terms for at least one pair $\epsilon_1, \epsilon_2$.

*Admissible estimates*

An estimate is *admissible* only if no other estimate is uniformly better than it. An estimator is admissible only if each of its estimates, appropriate to each body of data within its range, is admissible.

ADMISSIBLE ESTIMATES

These definitions are still in a vacuum. Everything hinges on how support for composite hypotheses is to be compared. Two cases seem easy. The first is that of simple dichotomy, when $d$ asserts that only two rival distributions are possible. Then we have no trouble with composite hypotheses and can rely directly on the law of likelihood. Or again, when $d$ permits use of the fiducial argument, we have ready-made measures of support for composite hypotheses, and these offer a forthright characterization of admissibility. But before exploring some consequences of these two special cases, it will be useful to compare our criterion with another theory on admissibility.

*Savage's criterion*

One widely acknowledged sign of superiority among estimators is given by Savage in a list of criteria for point estimation.[†] By now the properties of his criterion are quite well known. It is concerned with the average properties of estimators. An estimator is presumably better on the average than some rival estimator if it is closer on the average. But now arises the very same problem as for well-supported individual estimates. Consider two errors of different sizes, $\epsilon_1$ and $\epsilon_2$. An estimator $f$ might more often give estimates within $\epsilon_1$ of the true magnitude than the estimator $g$, and yet $g$ more often be within $\epsilon_2$ of the true magnitude than $f$. Which is 'closer on the average'? I think there is no definitely correct answer. But we can construct a very weak criterion for being closer on the average; it is exactly analogous to our earlier criterion for admissibility. Here we let $f$ and $g$ represent estimators; if the result $E$ is observed on some trial, $f(E)$ is the estimate made by $f$ in the light of that observed result.

We suppose as given in the initial data that $\Delta$ is the class of possible distributions. For convenience let a single parameter be under estimate; if the true distribution is $D$, represent the true value of this parameter by $\theta_D$. When is $f$ at least as close on the average as $g$? Certainly when for any error $\epsilon$, regardless of the actual distribution, the long run frequency with which $f$ gives estimates within $\epsilon$ of the true value of the parameter is at least as great as the frequency with which $g$ does.

It appears that we must consider the chance of a novel sort of

[†] L. J. Savage, *The Foundations of Statistics* (New York, 1954), p. 224.

event, the event that if $D$ is the actual distribution, an experimental result $E$ should occur such that
$$-\epsilon_1 \leqslant \theta_D - f(E) \leqslant \epsilon_2,$$
where $\epsilon_1$ and $\epsilon_2$ are positive errors. Represent this chance by
$$P_D(-\epsilon_1 \leqslant \theta_D - f(E) \leqslant \epsilon_2).$$
Savage asserted that if two estimators, $f$ and $g$, are such that for any pair of positive errors $\epsilon_1$ and $\epsilon_2$, and for every $D$ in $\Delta$,
$$P_D(-\epsilon_1 \leqslant \theta_D - f(E) \leqslant \epsilon_2) \geqslant P_D(-\epsilon_1 \leqslant \theta_D - g(E) \leqslant \epsilon_2)$$
with strict inequality between the main terms for some $D$, $\epsilon_1$ and $\epsilon_2$, then $f$ is better than $g$. Call this *being better in the sense of Savage's criterion*. We can all agree at once that if something is better in this sense, it will, if the initial data are correct, be closer on the average.

You might think that if one estimator is better than another, then any estimate made by the first would be more sensible than any made by the second. But not according to Savage's criterion. It is easy to discover a pair of estimators, $f$ and $g$, such that $f$ is better than $g$ (in the sense of Savage's criterion), yet for some particular observation $E$, $f(E)$ is a silly estimate while $g(E)$ is not.

| No. of heads in 5 tosses | $f$ | $g$ |
|---|---|---|
| 0 | 0·8 | 0·2 |
| 1 | 0·2 | 0·8 |
| 2 | 0·2 | 0·2 |
| 3 | 0·8 | 0·2 |
| 4 | 0·8 | 0·8 |
| 5 | 0·8 | 0·8 |

Take the simplest case of estimating $P$(Heads) from a coin and tossing device. It is known that $P(H) = 0·2$ or $P(H) = 0·8$; let further data concern the result of a sequence of five tosses. Let $f$ and $g$ take values given on the table above. $f$ is, by Savage's criterion, a better estimator than $g$, but $f$ tells us to estimate $P(H) = 0·8$ if the coin does not fall heads at all. This is pretty clearly foolish. $g$ gives an estimate of $P(H) = 0·2$, which would generally be acknowledged as more sensible. In fact the estimate of 0·8 on this data is not even admissible. Hence on Savage's criterion, $f$ can be 'better' than $g$, yet on some data $f$ can make a silly estimate while $g$ makes a good one.

This is just what we found for the Neyman–Pearson theory of testing. Distinguish two bodies of data, $d$ and $e$. $d$ shall state the class of possible distributions; in our case, that the chance of heads is either 0·2 or 0·8. $e$ shall state the result of a particular trial; in our case, that no heads were observed in our sequence of tosses. Now on data $d$, the before-trial data, the hypothesis that $f$ will give a closer estimate than $g$, is better supported than the contrary hypothesis that $g$ will be closer than $f$. So, if you care to express it this way, it is a better before-trial bet that $f$ will do better than $g$, than that $g$ will do better than $f$.

But now we make a trial and learn $e$; we should estimate the chance of heads on the basis of $d$ and $e$ taken together. We cannot pretend we do not know $e$. But on $d$ and $e$ together, the hypothesis that $f$'s estimate is better than $g$'s, is not better supported than the contrary hypothesis. In fact $d \& e$ shows that $f$'s estimate, in this particular case, is a foolish one, while $g$'s is not.

Thus we cannot regard Savage's criterion as automatically appropriate for after-trial estimation. At most it distinguishes good average estimators, and estimators good to bet on before conducting trials. Most of what was said in the discussion of Neyman–Pearson tests can now be repeated here.

However, in the example given, neither $f$ nor $g$ is what Savage would call a best estimator: other estimators are better, in Savage's sense, than both $f$ and $g$. It can easily be proved in the case of simple dichotomy that if an estimator is optimum in Savage's sense—such that no estimator is better in his sense—then it will be admissible. And all admissible estimators will be optimum in Savage's sense. So Savage's criterion agrees with that of admissibility in the case of simple dichotomy. This result is exactly parallel to a consequence of the Fundamental Lemma of Neyman and Pearson, that an optimum Neyman–Pearson test for simple dichotomy will always be a likelihood test, and vice versa.

*Use of the fiducial argument*

Simple dichotomy is of no general interest. We must now consider those problems for which degrees of support can be obtained from the frequency principle and the principle of irrelevance, that is, by the fiducial argument. This might at first seem a relatively narrow domain, but it provides for a surprisingly large part of classical estimation theory. It happens in the following way. For a given set-up the distribution of estimates made by most classical estimators is approximately Normal, and at least asymptotically approaches Normalcy. The actual distribution for an estimator will of course depend on the true value of the $\theta$ under estimate; the mean will be some known function of $\theta$, say $M(\theta)$. Now if the estimator has a Normal distribution, we know that the actual distribution of chances for different estimates is a member of the Normal family. So we may consider hypotheses about the actual distribution of chances for estimates. Since the distribution is supposed to be Normal, we can use the fiducial argument, as shown at the end of ch. IX. Hence we can get measures of support for different hypotheses about the location of the mean, $M(\theta)$. Since $\theta$ is functionally related to $M(\theta)$, hypotheses about the location of the mean are equivalent to hypotheses about the location of $\theta$. So we have obtained measures of support for composite hypotheses about the location of $\theta$. That is all we need in order to apply our criterion of admissibility among estimates. Thus from a class of rival estimators which are Normal or at least very nearly Normal, we can typically distinguish those estimators which are uniformly better than others.

*Unbiasedness*

Before attending more closely to approximately Normal estimators, it is useful to examine a special case. It may happen that in the long run the average of the estimates made by some estimator is exactly equal to the true value of what is under estimate. That is, the expectation of the estimate equals the true value. Such an estimator is called *unbiased*, which is unfortunate, for this usage of the word is not related to its application in the theory of testing.

It has quite often been proposed that estimators should be unbiased, or at any rate that the best estimators are in fact unbiased. The thesis is no longer as fashionable as it once was, probably because no good reason for it has ever been given. Notice that there is not only no reason for believing that, in general, an unbiased estimator will give a better individual estimate than some biased estimator. There is also no reason for believing that in general unbiased estimators are better on the average than biased ones. For an estimator can on the average be persistently awful, but as long as its errors are of opposite sign, it may still be unbiased, and have an average estimate equal to the true value.

Of course it might be true that the best of unbiased estimators were better than estimators of other kinds, even though inferior unbiased estimators are inferior to other estimators. But this has never been proved. Again it might be true that some very good estimators are unbiased, but this would be an incidental fact. We cannot use unbiasedness as a criterion of excellence. But it is still a useful concept for explaining other ideas.

*Minimum variance*

It so happens that for many estimators which people have actually considered, the distribution of estimates is at least approximately Normal, and if not actually Normal is at least a slightly warped bell-shaped curve. Moreover, many estimators so considered have in fact been unbiased. Our criterion of admissibility usefully distinguishes between such estimators. It is easy to check that among unbiased estimators with at least approximately Normal curves, those with smaller variance are uniformly better than those with large variance. In fact it can be proved that among unbiased bell-shaped estimators, if there is an estimator of minimum variance, then only that estimator is admissible. Hence among such estimators, minimum variance plays the crucial role. This conclusion follows directly from our very weak analysis of what it is for one individual estimate to be better than another. No facts about the average properties of estimators have been employed.

Minimum variance has had a long history in statistics, but few enough reasons have ever been given why it should pick out good estimators. Kendall's encyclopedia of advanced statistics succinctly reflects the common belief about variance: 'Generally speaking, the estimator with the smaller variance will be distributed more closely round the value $\theta$; this will certainly be so for distributions of the normal type. An unbiased consistent estimator with a smaller variance will therefore deviate less, on the average, from the true value than one with a larger variance. Hence we may reasonably regard it as better.'†

Even if we did accept this plausible argument for distinguishing among estimators, it would still be a question whether any individual estimate made with one estimator would be better supported

† M. G. Kendall and A. Stuart, *The Advanced Theory of Statistics* (3 vol. ed.) (London, 1961), II, p. 7.

than that made by another. So Kendall's argument would not hold for comparing individual estimates made by different estimators. Our theory shows that among unbiased estimators with approximately Normal curves, only that with minimum variance, if it exists, is admissible. That is, though this estimator is not necessarily best among all estimators, it is best among unbiased ones of roughly Normal form. If, indeed, lack of bias were a good criterion for estimation, then we should have an unequivocal theory for distinguishing, in some cases, uniquely best estimators. But lack of bias has never been proved a useful criterion for distinguishing good estimators from bad.

*Consistency*

Thus far we have considered estimators which can handle any outcome of a trial of some kind. But in estimating a parameter more and more trials can be made: a larger and larger sample obtained. So consider a sequence of estimators, the first member pertaining to outcomes of one trial of kind $K$, the second to outcomes of two consecutive trials of kind $K$ (namely, a compound trial), the third to three trials of that kind, and so on. It can be proved that if all are admissible estimators of a parameter with true value $\theta$ then if the mean of their estimates converges at all it will converge on $\theta$. This characteristic of such a sequence of estimators is called *consistency*.

Consistency has been deemed desirable for its own sake. Fisher, who made it central to the theory of estimation, called it 'the common-sense criterion' for estimation; at the end of his life he called it the 'fundamental criterion of estimation'.† Kendall accurately reports current opinion when he says, by way of introduction to consistent estimators, 'the possession of the property of increasing accuracy is evidently a very desirable one'.

Perhaps consistency is desirable: certainly it has been desired. It hardly indicates useful 'increasing accuracy', for an estimator, even on the average, may fail to be increasingly accurate over any humanly practicable number of experiments—and still be consistent. Yet an estimator can be 'increasingly accurate' over

† R. A. Fisher, 'On the mathematical foundations of theoretical statistics', *Philosophical Transactions of the Royal Society*, A, CCXXII (1922), 316; *Statistical Methods and Scientific Inference*, p. 141.

that human domain and still fail to be convergent. There is also a question of principle whether convergence should ever properly be taken to indicate increasing accuracy. But all these matters are academic. For any converging sequence of admissible estimators as described above is consistent.

*Maximum likelihood*

Maximum likelihood estimators give as the estimate of a parameter that value with the greatest likelihood; essentially, in our terminology, the value it takes on in the best supported hypothesis. Such estimators are admissible in our standard cases, that is, when there is simple dichotomy, or when the fiducial argument can be used. So much can be rigorously proved, but it does not settle the suitability of maximum likelihood methods for other problems.

Fisher defended his method in terms of its asymptotic properties. Maximum likelihood estimators are, first of all, consistent. The distribution of estimates in fact approaches an unbiased Normal curve of minimum variance.† So we have admissibility at the limit. Fisher believed these asymptotic properties crucial to the foundation of estimation theory, but this is not obviously correct. The properties may be, as it were, magical properties of admissible estimators, and not in themselves criteria of excellence. Fisher did, however, define a useful way of comparing the asymptotic properties of various estimators. Essentially, this rests on a comparison of their variance; the asymptotic unbiased minimum variance is compared to the variance of any other estimators; the ratio measures what Fisher called the efficiency of other estimates.

It is to be noted that only asymptotically do maximum likelihood estimators become unbiased and of minimum variance. In general they are not. They may always be admissible, but this does not mean they are best: only that nothing is uniformly better than they. It may happen that we have two admissible estimators, one maximum likelihood, and the other which asymptotically approaches an estimator less efficient than the asymptotic maximum likelihood one. But this does not seem an overwhelming reason for preferring the maximum likelihood estimator in finite domains.

† The general conditions for these asymptotic properties are in A. Wald, 'Note on the consistency of the maximum likelihood estimate' *Annals of Mathematical Statistics*, xx, 1949, 595–601.

*Minimax theory*

The minimax theory is properly part of the theory of decisions, in the form given to it by Wald's researches. We may, however, extract the central idea in terms of estimation: make estimates in such a way as to minimize the maximum possible expectation of error. To use the notion we require a measure of error. Gauss is usually followed, and the error of an estimate is measured by the square of the distance of the estimate from the value under estimate. In a richer decision theory the error would be weighted in terms of the costs resulting from it. Notice, however, that we will get different estimates according to how the error is measured. More generally we should consider an error function, which is a function of the distance of the estimate from the true value. A simple error function $E(\theta, \lambda)$ will be one such that, where $\theta$ is the true value, and $\lambda_1$ and $\lambda_2$ estimates,

$$E(\theta, \lambda_2) \geqslant E(\theta, \lambda_1)$$

for any $\theta \leqslant \lambda_1 \leqslant \lambda_2$ and any $\lambda_2 \leqslant \lambda_1 \leqslant \theta$.

Here we shall content ourselves with some very brief observations. First, for a wide range of problems with a simple error function the corresponding minimax estimator is in our sense admissible. In fact the error function generally picks out exactly one of the many admissible estimators, and is in this sense a solution to the problem of point estimation. Secondly, maximum likelihood estimators based on large samples approximate very closely to minimax estimators.

Actual minimax estimators are, from a traditional point of view, a trifle eccentric. Consider for instance the problem of estimating the chance of heads in independent tosses of a coin, on the data that the coin falls heads $k$ times in $n$ independent tosses. A common-sense, a maximum likelihood, a minimum variance, an unbiased, and indeed a method of moments estimate is $k/n$. But using Gauss' error function, the minimax method gives,

$$\frac{1}{1+\sqrt{n}} \left( \frac{k}{\sqrt{n}} + \frac{1}{2} \right). \dagger$$

† J. L. Hodges and E. L. Lehman, 'Some problems in minimax point estimation', *Annals of Mathematical Statistics*, XXI (1950), 190.

Carnap protests vigorously. He believes that if $a$ and $b$ add up to $c$, and $c$ is known, then the estimate of $a$ plus the estimate of $b$ should add up to $c$.† Now we know the chance of heads plus the chance of tails must add up to 1; but their minimax estimates do not add up to 1. So Carnap feels strongly inclined to reject the estimates. Everyone must be surprised at the minimax result, but although it may at first seem counter-intuitive on further reflexion it seems impossible to state any substantial reason for demanding additivity. I am not sure, but I suspect there is no good reason why estimates should be additive, and hence, on this score, no sound ground for rejecting the minimax estimate.

Fisher, as was remarked in the preceding chapter, held that estimates are intended as summaries of statistical data, and hence that an estimate of $\theta$ is also an estimate of the value of every single-valued function of $\theta$. Hence he would reject minimax estimates out of hand. But on our account of an estimate, where an estimate is something which aims at being close to the true value, we cannot follow his course.

## Conclusion

If by an estimate is understood something which aims at being close to the true value of what is under estimate, there are in general no uniquely best estimates. This is because of an indeterminacy in the notion of an estimate. It is not so much because, in Carnap's words, there is a 'continuum of inductive methods' but because the concept of point-estimation is not well adapted to statistical problems. Nor is the matter resolved by transferring attention from individual estimates to average properties of estimators, for the average properties which have been claimed as desiderata typically lead to estimators which, in our sense, are admissible; they do not, with any justice, sharpen that class.

What to do? The first possibility is to ignore the matter altogether. On the mathematical side, develop the general theory of admissible estimators, and hope to discover new plausible criteria of interest or elegance. On the practical side, observe that in most problems assuming a fair amount of data, different admissible estimates are generally rather close together. Hence use an estimator which is convenient for computation. If estimates are really

† R. Carnap, *The Continuum of Inductive Methods* (Chicago, 1952), pp. 81–90.

necessary for practical decisions based on small amounts of data, by-pass estimation and find what comfort decision theory can offer.

The second possibility is to decree, by convention, that some particular method of estimation will be called the best. To many readers the method of maximum likelihood will be attractive. It is completely general, applying to almost any practical situation which will arise. For smooth likelihood functions its estimates are usually and may be always admissible. Even when it is difficult in practice to compute the maximum likelihood estimate for some kinds of data, there are useful approximations to this estimate which are pretty easy to compute. Decision theorists take comfort in its asymptotic properties. And it may seem nice to give as one's estimate what is best supported by the data. But all these are at most desiderata inclining one to adopt a convention. They in no way show that, as far as the present meaning of the word 'estimate' goes, maximum likelihood estimates are unequivocally best.

A third possibility is to abandon the whole conception of estimation in connexion with statistics, although one might revert to Fisher's idea of estimates, together with ancillary statistics, serving as very brief summaries of data. For more practical affairs one would replace discovering best estimates by discovering directly what, if any, are the best acts to make on the basis of some data and some knowledge about the consequences of various acts.

A fourth possibility is to introduce some completely new concept into the study of chance, and try to analyse estimation in the light of results which it discovers. This must occupy us for two chapters; in ch. XII we shall examine the fundamental new idea of Thomas Bayes, and then in XIII, develop the theory of his successors.

A fifth possibility is to try to refine the logical theory of support. For defects in the concept of estimation are not entirely responsible for the limitations in our results. No matter what be meant by support, nor how it be analysed, there will still be difficulties, of the sort mentioned on p. 177 above. But at present we are hampered, for although we can tell, on our theory, whether one statistical hypothesis is better supported than another, we cannot in general tell whether a disjunction of statistical hypotheses is better supported than another disjunction; still less can we tell whether one infinite set is better supported than another. It would be nice to overcome this by a completely general method for measuring or at least comparing the support for sets of statistical hypotheses. If we could accomplish this we could certainly strengthen our definition of uniform superiority among estimates, and achieve a far tighter definition of admissibility.

CHAPTER XII

# BAYES' THEORY

The physical property of chance was defined, or partly defined, in the logic of support. This definition has been applied both to testing hypotheses and to making estimates, but however pleasing some individual results, solutions to the most general problems have become increasingly meagre. Much of our work applying the theory has been piecemeal and empirical. In contrast, several theories aim at explaining all the principles of inference in one swoop. If any is correct, we shall have to embed chance and support within it. I do not believe any is entirely correct, but it is worth stating why each seems doubtful; the theories already have many critics, but it may be useful to discuss them from the point of view of our theory of support.

The various rival theories could be presented at once in modern format, but it is instructive to take some semblance of an historical approach. The name of Thomas Bayes will recur again and again. From the beginning we must distinguish his two distinct contributions. One is a postulate, the central subject of this chapter. The other is a theorem, a consequence of the Kolmogoroff axioms. We have already noticed it in discussing measures of support. He proved that,

$$P(A|BC) \text{ is proportional to } P(B|AC).P(A|C)$$

and the corresponding theorem for densities. The constant of proportionality depends on the $B$ and $C$, but is constant for all $A$. The proof is quite elementary, and I am not certain that Bayes was the first to see it. The theorem would never have been singled out were it not for its applications; it is called after Bayes because he was the first to use it in arguing for a method of statistical inference.

Being a consequence of Kolmogoroff's axioms, the theorem holds for any interpretation of those axioms. In particular, it holds if '$P(\ |\ )$' be interpreted as a mere conditional chance, or long run frequency. It holds when $A$, $B$, $C$ are assigned propositions and '$P(\ |\ )$' is interpreted as the degree to which one proposition supports another. Finally, to foreshadow developments in this chapter, it holds if '$P(\ |\ )$' is interpreted as the fair odds for betting on one event, given knowledge of another.

The theorem is easily remembered if the second or third interpretation is kept in mind. Take the second. The theorem says that $P(A|BC)$ is proportional to $P(B|AC)(P(A|C)$. Let $A$, $B$ and $C$ be propositions; more particularly, let $A$ state an hypothesis, $C$ some initial knowledge for assessing the hypothesis and $B$ some further data, perhaps a new experimental result. Then $P(A|BC)$ is the degree to which $B$ and $C$ support $A$: the support for $A$ in the light of both old and new data. If $B$ is the result of a new experiment, $P(A|BC)$ may be called the *posterior support* for $A$, that is, the support available for $A$ after experiment $B$. In this vein, $P(A|C)$ is the *prior support* for $A$, the support for $A$ before learning of $B$. Finally, $P(B|AC)$ is the support for the experimental result $B$ on hypothesis $A$ coupled with data $C$; this corresponds to the chance of getting $B$ if $A$ is true, and so to the *likelihood* of $A$ in the light of $B$. Thus Bayes' theorem may be remembered as stating,

*posterior support* $\propto$ *likelihood* $\times$ *prior support*.

Precisely the same mnemonic applies in terms of chances or betting rates. Or letting $P$ be uninterpreted, the theorem says that the posterior $P$ is in proportion to the product of the prior $P$ and the likelihood.

## Bayes' definitions

There is no more fitting introduction to uses of the theorem than Bayes' celebrated memoir.† He first solves a problem in the familiar doctrine of chances—in the computation of new chances from old—and then uses this to argue for a radical new postulate which, if true, would make all our work on the logic of support look like the halting progress of a blind beggar. Before explaining the postulate we must examine a new concept. Then we shall see how it applies to a straightforward problem, and finally study the analogical argument for the new postulate. To begin with his crucial definitions:

† T. Bayes, 'An essay towards solving a problem in the doctrine of chances', *Philosophical Transactions of the Royal Society*, LIII (1763), 370–418. Reprinted in *Biometrika*, XLV (1958), 296–315.

5. The *probability* of any event is the ratio between the value at which an expectation depending upon the happening of the event ought to be computed, and the value of the thing expected upon its happening.

6. By *chance* I mean the same as probability.

In communicating the paper to the Royal Society, Price, its sponsor, reports Bayes as feeling apologetic about these entries but adds, 'instead of the proper sense of the word *probability* he has given that which all will allow to be its proper measure in every case where the word is used'.

'The value of the thing expected upon its happening' may usefully be compared to the value of a *prize* won when the event in question does occur. 'The value at which an expectation depending on the happening of the event ought to be computed' is more obscure. It seems correct to construe it as the *fair stake*. If one must bet on the occurrence of an event, the fair stake is what ought to be wagered in order to win the prize; he wins the prize if the event does occur, and loses the fair stake if it does not. Bayes mentions neither prize nor fair stake, but all that he writes is compatible with use of these terms.

*Some assumptions*

The notion of a fair stake is rife with difficulty, but for the moment we shall take it for granted and see where it leads. A few assumptions are needed to explain Bayes' theory of inference. It will be noted that whether a stake is fair or not must depend in part upon the evidence available to bettor and bookie where, to be fair, we suppose both persons share their knowledge and skills. When an event is known to have happened or when it is certain that it will happen, the fair stake must equal the prize; when the event is not known, the stake should be less; if the event is known not to have happened, the only fair stake is 0. So we tentatively assume:

1. For given prize and given event, the fair stake depends on the available information concerning the occurrence of the event.

Bayes evidently assumes that for given information the ratio between the stake and the prize is independent of the size of the prize. This is, perhaps, a substantial assumption about some scale of value, say money, or, in a more modern view, is a stipulation to be used in defining a scale on which the utilities of both stake and prize are to be measured. Anyway, we have:

## SOME ASSUMPTIONS

2. Let there be a given collection of information, and a possible event. If there is a fair stake for wagering on the occurrence of the event in order to win a positive prize, the ratio between stake and prize is independent of the size of the prize.

This ratio is the *fair betting rate*; it defines the fair odds in betting for and against the happening of some event. It closely resembles, and may be identical with, what Bayes defined as the 'probability'. Many will protest that naming something a fair betting rate is mere hypostatization, but here we grant what Bayes seems to assume, in order to see where it leads. He also seems to assume that betting rates should equal chances in accord with what some chapters ago, I called the frequency principle. More exactly:

3. When the chance (long run frequency) of getting outcome $E$ on some trial of kind $K$ from some set-up $X$ is known to be $p$, and when this is all that is known about the occurrence of $E$ on that trial, then the fair rate for betting on $E$ should be $p:1-p$.

Perhaps Bayes made no such assumption; he does not allude explicitly to a physical property of chance, but he writes in such a way that it is convenient, in expounding his fundamental ideas, to suppose that he had something like (3) in mind. I must warn, in aside, that although many people readily grant (3), it is by no means certainly correct.

Finally Bayes argues that, as we might express it now:

4. In a given state of information, betting rates satisfy Kolmogoroff's axioms where '$P(E)$' in those axioms is interpreted as the fair rate for betting on $E$, in the given state of information.

In his paper Bayes uses some considerations about fair betting to establish (4); in the next chapter I shall describe some simpler arguments which have recently been invented.

### Subjective and objective

In assumption (1) I have concealed a superficial difficulty in reading Bayes. Sometimes he writes as if the fair betting rate is entirely a function of the available information, and may alter as any new information is made available. At other places he is at odds with this idea; he writes of the unknown probability of an event as if there were an objective property of it quite independent of whatever information is available.

Although he is not explicit on this issue, there is no reason to think he is confused. For although, as said in (1), the betting rate may depend on what information is available, the dependence may not be complete. It is useful to contrast two gambles. First a horse race: there is a strong inclination to say that if there is a fair rate for betting on some particular horse at one definite race, the rate is entirely a function of what is known, and should change, say, when it is learned that the favourite has been scratched. But take a bet at roulette; suppose at the beginning of play it is universally agreed that the fair rate for betting on red is just under one half; suppose this accords with everything known to bettors and croupier. If, however, it turns out that the chance, the long run frequency, of red is a mere 1/4, there is a strong inclination to say, not that the fair betting rate has changed, but that the rate all along should have been 1/4. It is tempting to say that 1/4 is 'really' the fair rate, and was all along; at first it was simply unknown.

Some readers will have been trained to discard one of the two inclinations I mention. Others will suppose that their very existence betrays two different meanings of 'fair' or of 'fair betting rate'. I guess, on the contrary, that a correct analysis of the meaning of the English word 'fair' would show only one sense in this connexion, and accord with all that has been said. But here the analysis would be too long and perhaps too devious. Bayes' ideas can be expounded without such subtleties. All the 'probabilities' Bayes considers fall pretty clearly into two classes—those which are entirely 'objective' and which may be unknown, and those which are entirely relative to the available evidence. On any occasion where it matters he makes plain which class is in question. So I shall explain his work by using 'fair betting rate' as an entirely relative notion.

In this usage, the fair betting rate on red is 1/4 only if available information shows it to be 1/4. So when the probability mentioned in one of Bayes' problems is evidence-dependent, I construe it as a fair betting rate. When it is objective I construe it as equal to the chance, or long run frequency. Bayes has written with such clarity that this can be done without falsifying any of his ideas. Moreover, nothing I have said implies that Bayes used the word 'probability' in several different senses. It is perfectly possible that what Bayes called probability was such that the probability of

## SUBJECTIVE AND OBJECTIVE

some things is evidence-dependent and of other things is not. Indeed his ideas can, as here, be explained using two different terms; because of recent controversy this is surely advisable; but it does not follow that his original usage was in any way ambiguous. I could explain an early meteorological treatise by using one term for powder snow, and something etymologically unrelated for snow which falls in large fluffy flakes, but it would not follow that the word 'snow' has two different meanings.

### Bayes' problem

'*Given* the number of times in which an unknown event has happened and failed: *Required* the chance that the probability of its happening in a single trial lies somewhere between any two degrees of probability that can be named.' Judging by his treatment, this problem should be explicated in my terms thus: *Given* a chance set-up $X$ and a kind of trial $K$, and given the number of trials of kind $K$ on which $E$ has occurred, and the number on which it has failed to occur: *Required* the fair betting rate, based on the given information, that the chance, or long run frequency, of the outcome in question lies between any two degrees which can be named.

Bayes first solves this problem for what might be called *chance set-ups in tandem*: a pair of set-ups such that the distribution of chances on the second is itself determined by the outcome of a trial on the first. For instance, the distribution of chances of various crop yields might be a function of the year's weather, itself conceived as the outcome of a trial on a set-up. Or perhaps there is a distribution of chances of a man's staking various sums at the gaming table; this distribution may depend on his luck the previous night.

When a distribution of a set-up $X$ is itself the outcome of a trial on another set-up $Y$, $Y$ shall be called the *parent* set-up, and $X$ the *junior*. To understand the problem Bayes set himself, take the simplest kind of example. A botanist wants to know the rate at which cells of type $A$ travel down the stem of a plant. He labels a cell with radioactive iodine and plots its progress. But there are two complications. First, there are two kinds of cell he might have labelled, $A$ and $B$. He cannot tell which he labelled without ruining his experiment. But he knows the chance of labelling an

$A$ is 0·4, and a $B$ is 0·6. Secondly, a labelled cell may either pass through a certain node in the stem, or may be stopped there. The chance of a $B$ being stopped is 0·7, and of an $A$ is 0·3. His labelled cell does pass through the node. What is the fair rate for betting that it is an $A$?

We have the parent set-up: there are two possible results, 'labelling an $A$' and 'labelling a $B$'. We have the junior set-up: two possible results, 'cell passes the node' and 'cell is stopped'. However, we can also treat this as a single set-up, thus regarding the two set-ups as operating in tandem. Then there are four possible results:

(1) an $A$ is labelled, and it stops;
(2) a $B$ is labelled, and it stops;
(3) an $A$ is labelled and it is not stopped;
(4) a $B$ is labelled and it is not stopped.

The chances of these four events can be computed from their parts; the chance of (3) is 0·4 × 0·7, namely 0·28. Moreover, one can compute conditional chances. For instance, the chance of a labelled cell being $A$, given that the cell was not stopped, is just

$$P(A|\text{not stopped}) = \frac{P(A \& \text{not stopped})}{P(\text{not stopped})}$$

$$= \frac{P(3)}{P(3 \text{ or } 4)} = \frac{0·28}{0·28 + 0·18} = \frac{14}{23}.$$

So the chance of getting $A$, given that the cell did not stop, is rather better than a half. Since fair betting rates are, by one of our assumptions, taken equal to chances, we may infer that 14/23 is the fair betting rate that the cell is $A$, given the initial information including the fact that the cell did not stop. Or this problem could be solved using Bayes' theorem for chances.

It is notable that the whole calculation, until the final step, may be conducted in terms of chance. Betting rates come after the arithmetic, and could, for purposes of inference, be entirely ignored. We could, for instance, use our logic of support to infer from the computed chances, which of several hypotheses is best supported. This is just the logic of chance and support. So far, there is nothing peculiar about Bayes' work; it is entirely uncontroversial. It is the theory of tandem set-ups.

*Bayes' own example*

Bayes' own example is elegant, and will be useful for following a later stage in his argument. Copying his notation and his diagram, there is a square table $ABCD$ so made and levelled that if either of the balls $O$ or $W$ is thrown upon it, the chance that it shall come to rest on any part of the square is equal to the chance of its resting on any other part. Moreover, either ball must come to rest somewhere in the square.

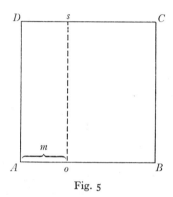

Fig. 5

For a trial on the *parent set-up*, toss in $W$ and draw, through the point where it comes to rest, a line *os* parallel to $AD$.

For a *simple trial* on the *junior set-up resulting from a trial on the parent*, toss in ball $O$. If $O$ rests between *os* and $AD$ then, whatever the location of *o*, we say the event $M$ has occurred. A *compound* trial on the junior set-up consists of noting the number of occurrences of $M$ in $n$ simple trials.

Evidently the chance of $M$ on any simple trial on a junior set-up is determined by the outcome of the preceding trial on the parent set-up. In fact, if the table is a unit square, and on the parent trial the line *os* comes out at a distance $m$ from $AD$, the chance of $M$ is just $m$.

Bayes' problem: if the outcome of a trial on the parent set-up is not known, but a compound trial has result $k$, ($k$ out of $n$ occurrences of $M$), find the fair rates for betting that $P(M)$ lies between any two degrees of chance, $p$ and $q$. (Or, equivalently, find the betting rate that the line *os* lies between distances $p$ and $q$ from $AD$.)

The solution proceeds just as in the elementary example. No need to do the computation here; suffice to be sure that we treat the two different trials as a single trial on a tandem set-up. No use need be made of betting rates while computing, though a betting rate may be inferred from a computed long run frequency. If you want, you can use Bayes' theorem to render the calculation entirely routine.

*Bayes' postulate*

After this unexceptionable work on tandem set-ups, Bayes turns to the case in which it is *not* known that the chance set-up of interest is itself determined by the outcome of another, parent, set-up with known distribution. He is concerned with some chance set-up $X$; he wants to know the chance of getting outcome $E$ on a trial on the set-up. He has *initial data*, stating the class of possible distributions for the set-up $X$, and hence stating the possible values of $P(E)$. He considers only cases in which $P(E)$ may, for all the initial data say, lie anywhere between 0 and 1.

Every possible value of $P(E)$ may be conceived as an event—the event that $P(E)$ has such and such a value. It seems to make sense to bet on the true value of $P(E)$, or that $P(E)$ lies between $p$ and $q$. Bayes postulates a *uniform distribution of betting rates* over all possible values of $P(E)$; it is the distribution such that the fair rate for betting that $P(E)$ lies between any two degrees of chance, $p$ and $q$, is equal to $|p-q|$. Since by assumption (4) made earlier in the chapter, this distribution will satisfy Kolmogoroff's axioms, this distribution is unique, aside from points where $P(E) = 0$. Thus Bayes postulates a virtually unique distribution of betting rates, given what I have called his initial data.

It will be observed that the uniform distribution has exactly the same formal properties as a distribution of chances. Moreover, suppose we learn the outcomes of some trial or sequence of trials on a set-up $X$. Then, on this new data, it will be possible to compute by Bayes' theorem the fair betting rate that $P(E)$ lies between any two degrees of chance, $p$ and $q$. This calculation will be, formally, exactly the same as for the tandem set-up involving the billiard table. Its significance is, of course, totally different, but the calculation, and the numerical answers for the betting rates, will be just the same.

Bayes' postulate solves more than the problem with which he began. Let a class of hypotheses be up for test. Does the true hypothesis lie in this class? Bayes can first take a uniform distribution of betting rates and then, on further experimental data, compute the betting rates appropriate to this more substantial data. If it is a good bet that the class of hypotheses includes the true one, accept the class; if it is a good bet that the class does not include the truth, reject it. Or suppose it is a problem to compare two interval estimates. Compare the betting rates that these include the true distribution. Whole new vistas of inference open up before us. So we had best see if Bayes' postulate is true.

*Bayes' justification*

Bayes knew his postulate could not be proved from his definitions, but defends it in a famous scholium. Consider any chance set-up $X$, and any event $E$, when there is no initial information about the location of $P(E)$. 'Concerning such an event', writes Bayes, 'I have no reason to think that, in a certain number of trials, it should rather happen any one possible number of times rather than another. For, on this account, I may justly reason as if its probability had been at first unfixed, and then determined in such a manner as to give me no reason to think that, in a certain number of trials, it should happen any one number of times rather than another. But this is exactly the case of the event $M$.' At this juncture he recalls the tandem set-up in which $P(M)$ is determined by the location of a point $o$, in turn determined by the fall of a ball $W$. Completing his comparison of $M$ with $E$, he says of $M$ that 'before the place of the point $o$ is discovered or the number of times the event $M$ has happened in $n$ trials, I have no reason to think it should happen one possible number of times rather than another'.

I fancy Bayes argues: (i) Before any trials on the billiard table, and before the point $o$ is discovered, there is no reason to suppose $M$ will happen any number of times rather than any other possible number—and, he might have added, there is no reason to prefer any value of $P(M)$ to any other. (ii) Exactly the same is true of the event $E$, in the case that no parent set-up is known. (iii) Betting rates should be a function of the available data: when all the information in the two situations is formally identical, the betting rates must be identical. (iv) In all that matters, the data in the case of $E$ and $M$ are identical. (v) The initial distribution of betting rates for $P(M)$ is uniform: it assigns equal rates to equal intervals of possible values of $P(M)$. Therefore, (vi) this should also be the initial distribution of betting rates for $P(E)$.

It is not altogether easy to pin-point the fallacy. We must put a dilemma, connected with the reason for assertion (v). *Interpretation A*: (v) does not follow from (i) directly, but is the consequence of the fact that the table is so made and levelled, that the long run frequency with which the ball falls in any area is equal to the long run frequency with which it falls in any other area; we infer (v) from this fact plus assumption (3). *Interpretation B*: (v) does follow from (i) directly.

Most readers since the time of Laplace have favoured $B$, but I think there is every reason to suppose Bayes intended something like $A$. Why else, both in his paper and his scholium, was he at such pains to compare $E$ to $M$? For if (v) follows from (i) directly, then (vi) follows from (ii) directly, without any need of mentioning $M$ at all. Because he so persistently argues by the analogy between $E$ and $M$, we must suppose he had $A$ in mind. But under interpretation $A$, the argument is fallacious. For if assumption (3) and the facts about frequency are needed to obtain (v), then we must use data about $M$ which are not available for $E$, and so assertion (iv) is false. Hence the demonstration is invalid because it uses a faulty premiss, namely (iv).

Interpretation $B$, on the other hand, is fallacious because it employs a faulty inference from (i) to (v). In $B$, it is held that mere lack of reason for supposing $P(M)$ lies in one short interval rather than another of the same size entails the betting rate on equal intervals is in proportion to their size. Hence the betting rate that $P(M)$ is less than 0·5 should be 1/2. But to take Fisher's well-known counter-argument, there is equally, on the given data, no reason for supposing that the value of arc sin $P(M)$ lies in any interval between 0° and 90°, rather than any other of the same angular size. So by parity of reasoning the betting rates should be proportional to angular size. So the betting rate that arc sin $P(M)$ is less than 30° should be 1/3. But the event that $P(M)$ is less than 0·5 is identical to the event that arc sin $P(M)$ is less than 30°. Hence this event has betting rates 1/2 and 1/3 on the same data.

By taking more esoteric transformations than arc sin we may compute any betting rate we please. So interpretation $B$ will not do.

Cautious Bayes refrained from publishing his paper; his wise executors made it known after his death. It is rather generally believed that he did not publish because he distrusted his postulate, and thought his scholium defective. If so he was correct. To judge by a number of commentaries, modern scholarship imagines Bayes intended the scholium in interpretation $B$. If so, it is but one more version of the tedious principle of indifference. Yet there is little reason to suppose Bayes had this in mind. What he says has little to do with $B$, and much to do with tandem set-ups. So I suppose he must have meant something like $A$, though possibly that does not entirely capture his meaning either. However, since there is no known way of constructing Bayes' argument so that it is valid, there is no reason to suppose Bayes' postulate true.

*Jeffreys' theory*

Forms of Bayes' postulate have not lacked defenders, one of whom is incomparably finer than any other. Jeffreys' *Theory of Probability* is the most profound treatise on statistics from a logical point of view which has been published in this century. It marks the first attempt to apply the axiomatic procedures, familiar to logicians, in deriving, criticizing, and contributing to the methods of statisticians. For long there had been axiomatic calculi corresponding to what I call Kolmogoroff's axioms, but none stated principles of statistical inference as so many further axioms. I think the debt to Jeffreys in preceding chapters of this book is obvious. What I call the logic of support is no more than a form of what he calls the logic of probability. But here we are concerned solely with one aspect of his work: his use of modified Bayesian postulates as a foundation for statistical inference. Only this part of his foundational work will here be under scrutiny.

Like so many students of the subject I have had to conclude by dissenting from one of Jeffrey's key assumptions. If his work were properly appreciated, that would end the matter: nobody ever apologized to Bayes about rejecting his postulate, nor to Euclid on qualms about the fifth postulate, or to Frege because of Russell's paradox, or to Whitehead and Russell because the axiom of infinity is empirical and the axiom of reducibility either redundant or false. Yet much of Jeffreys' work has been largely ignored because of the idea that if one of a man's axioms is dubious then the rest of his work is useless. This is utterly to misconceive the task of foundations. Fifty years hence people will still be quarrying Jeffreys for what they think are pretty interesting boulders, which turn out to be elaborate pyramids. In the next chapter, on subjective statistics, we shall repeatedly be called up short because subjectivists have given some problem no philosophical investigation; in every case we can refer to Jeffreys' perspicuous analysis of the same difficulty. Yet in every case, what is important in Jeffreys is independent of the dubious postulates. To these we must here turn our attention.

Jeffreys' theory is developed in terms of what he calls probability, and which he explains as the degree of confidence which can be placed in a proposition on given data. The rules governing it are the rules for comparing degrees of just confidence, or, as he often puts it, of degrees of reasonable belief. He makes plain that his 'probability' is a relative notion; strictly speaking one must not speak of probability *tout court* but rather of the probability relative to such and such information.

He deliberately avoids following Bayes, and declines to present his theory in terms of fair betting rates. He has excellent reason: he has in mind the priority of the idea of confidence over the notion of a reasonable bet, the severe difficulties in measuring betting rates, and problems about a scale of utility. But because of formal similarities, it does not matter much for present purposes whether one considers his theory from the point of view of fair betting rates, or of degrees of support; we shall use his word, 'probability', continuing of course to use 'chance' to refer to long run frequency.

In the name of probability Jeffreys makes all the well-known assumptions which were numbered (1) through (4) at the beginning of this chapter; he also assumes, in effect:

5. For any possible event there exists, for any state of information, a unique probability for that event.

There are two parts to Jeffreys' theory. First there is a critical assignment of distributions of probability when virtually nothing is known except the class of possible events. Bayes' postulate suggests one possible assignment; generally Jeffreys prefers slightly different forms. These distributions are called *initial* since they fit the most meagre initial information; postulates stating initial distributions will be called *initial postulates*. Evidently many initial postulates are alternative to that of Bayes.

Given the initial postulates the remainder of Jeffreys' theory is seldom controversial. Posterior distributions are computed from prior ones by Bayes' theorem. Since our concern here is foundations, we must attend primarily to the initial postulates. Doubt about these has prevented most students from subscribing to Jeffreys' theory. It has been widely assumed that his postulates are just forms of Bayes' postulate, and should be rejected without more ado.

It certainly seems there is no way to choose initial distributions. So they seem pretty arbitrary. But although they cannot be uniquely determined, Jeffreys does not think this fatal to his theory. He analyses what the initial distributions are supposed to signify. Suppose we know only a class of possible statistical hypotheses, and nothing else whatsoever. Then, says Jeffreys, we are virtually ignorant. An initial distribution serves only as a mathematical expression of our ignorance. Choose what is mathematically most convenient, and what is at least consistent with our vague background of experience. In particular choose distributions which are invariant under all the most plausible transformations on methods of measurement. In some problems Bayes' uniform distributions are easiest to handle, but Jeffreys deplores the weight of the authority of Bayes and Laplace, and the idea that 'the uniform distribution was a final statement for any problems whatsoever'.† In different cases he postulates different distributions.

But in justifying his postulates it will not do simply to claim that the initial distributions are mere convenient -summations of ignorance. For they assert a distribution of probabilities, presumably also of fair betting rates and of degrees of support. According to Jeffreys' conception there is only one right distribution. So a particular assignment is either true or false. But there is in fact no reason to suppose that, in general, Jeffreys' postulates state the correct distributions.

The matter is made even more baffling by the fact that some of

† H. Jeffreys, *Theory of Probability*, p. 98.

the initial distributions and all the posterior distributions which Jeffreys recommends can be normalized; that is, we can express the degrees of probability by numbers no greater than 1. But the most interesting prior distributions cannot be normalized; degrees of probability range up to infinity. Now it is of course a matter of convention on what scale you choose to measure probabilities. But to use measurement on one scale to provide measurement on another—that is rather mysterious. If we have an unknown parameter which can range anywhere from 0 to $\infty$, we are usually told to assume that the prior probability of the logarithm of the parameter is uniformly distributed. Jeffreys argues persuasively that this represents our ignorance about the parameter. So we assent to probabilities that do not sum to any finite quantity. We substitute these in a formula, use some other data, and get probabilities which do sum to 1. What is going on here? It looks like magic; what, in the eighteenth century, was called formalism. Agreed, you can measure probabilities on any scale you please, but how can one pretend to understand a particular probability if this witchcraft moves us from one scale to another?

If the theory of support is accepted, together with the fiducial argument given in ch. IX, then in special cases it is possible to prove the existence of unique initial distributions, and if we were to admit distributions which do not normalize, these seem always to coincide with what Jeffreys wants. As was earlier remarked, his brief analysis of the fiducial argument is more sound than any hitherto published; he also says that, in effect, it is merely the inverse of his own kind of reasoning. But although the fiducial argument can be held to give striking confirmation of some of Jeffreys' prior distributions, it does not entail all of them, and in continuous cases provides some of them only if we admit his altogether unexplained switching of scales. It seems more reasonable to guess that very seldom are there uniquely correct initial betting rates or degrees of confidence appropriate to virtual ignorance. Humans are not in that primal state, but it hardly looks as if in such a state there could always be a uniquely correct distribution of probabilities. Yet even if there were such, how should we ever find it out? Concerned with the foundations of our discipline, we presumably want the true distributions, the true postulates.

Jeffreys seems to suppose that the true location of initial distributions is a purely logical problem, and that 'there is some reason to suppose that it would be solved fairly quickly if as much ingenuity was applied to it as is applied to other branches of logic'.† But in truth, despite Jeffreys' investigations, he does not seem to pose a clearly defined logical problem at all. Moreover, he is oddly inclined to say that for many applications the solution of the 'logical problem' does not matter much; 'Where there are many relevant data or where a crucial test is possible, the posterior probabilities are affected very little even by quite considerable changes in the prior probabilities; a course of action would be affected by such changes only in circumstances where the ordinary scientific comment would be "get some more observations". We can make considerable progress while leaving some latitude in the choice of initial probabilities.'‡ Unfortunately we are concerned with a question in logical foundations, and with whether some postulates are true. The fact that it would not matter much for some problems if the postulates were false is no defence of the truth of the postulates.

Jeffreys makes great play with what he calls a 'working rule given by Whitehead and Russell: when there is a choice of axioms, we should choose the one that enables the greatest number of inferences to be drawn'.§ Perhaps he refers to the following maxim: 'The reason for accepting an axiom, as for accepting any other proposition, is always largely inductive, namely that many propositions which are nearly indubitable can be deduced from it, and that no equally plausible way is known by which these propositions could be true if the axiom were false, and nothing which is probably false can be deduced from it'.∥ A good maxim, though Russell coined it to bolster the specious (or anyway notorious) axiom of reducibility. But the authors of *Principia* do not advise, as Jeffreys suggests, the axioms from which one can draw the greatest number of inferences, but the axioms from which one can draw the greatest number of true inferences without validating any which are probably false. So the maxim counts rather against Jeffreys' theory. Either there are uniquely correct initial distribu-

† H. Jeffreys, *Scientific Inference*, 2nd ed. (Cambridge, 1957), p. 33.
‡ Ibid. pp. 39 f.
§ Ibid. p. 33.
∥ A. N. Whitehead and B. Russell, *Principia Mathematica*, 2nd ed. (Cambridge, 1925), p. 59.

tions or there are not. If there are not, his postulates are false. But if there are uniquely correct ones, none we postulate as correct are likely to be exactly true; at any rate there is no reason, not even in deductive fruitfulness, for choosing one from a whole infinity of similar but incompatible postulates.

I do not think that Jeffreys' theory provides a proper foundation for statistical inference, but this is not to say that prior distributions are useless. The next chapter will describe one kind of application based on subjective considerations; but that is alien to Jeffreys' ideas. However, suppose we do have a collection of observations, and we take some initial distribution which is pretty flat in the region around those observations. Any more or less flat distribution would do. Then if we compute the posterior distribution from the prior one, in accord with Bayes' theorem, we presumably gain some understanding of what the data do to a flat distribution. And this, I take it, is one way of getting some insight into the data. Here the flat distribution may be conceived, in Jeffreys' way, as some kind of expression of ignorance. There is no need for it to be total ignorance. Both in Jeffreys own work, and also in a recent paper,† there are examples where the initial distribution for study is chosen with some regard to the magnitude of the actual observations, and so does not record so much total ignorance, as a knowledge based on a first, cursory, view of the observations. This is, I believe, important for gaining a qualitative view of the data, where we are concerned with what data do to any more or less flat distribution.

## *Tu quoque?*

I imagine it may now be retorted that my theory of statistical support is guilty of the same assumptions as that of Jeffreys and Bayes. It will be argued that, first, when the statistical data merely give the possible distributions, all simple statistical hypotheses are equally well supported, and that this is tantamount to a uniform distribution of betting rates on initial data. Secondly, most inferences permitted by my theory coincide with ones allowed in Bayes' theory; this should confirm the suspicion that assumptions about prior probabilities are illicitly concealed in the theory of

† G. E. P. Box and D. R. Cox, 'An analysis of transformations', *Journal of the Royal Statistical Society*, XXVI (1964).

support. A Bayesian would instruct us to admit this openly, and adopt a theory of initial distributions; an anti-Bayesian would condemn our theory forthwith.

These objections are utterly unfounded. Contrary to what is suggested, the equal support of statistical hypotheses on minimal statistical data neither entails nor is entailed by a uniform distribution of betting rates. It does not entail it, for nothing in the logic of support suggests that equal support for some sets of distributions follows from equal support for the members of those sets; yet the uniform distribution theory is explicitly a theory about sets of distributions. Nor is our theory entailed by the Bayesian theory. Take the case of continuously many possible distributions. Then on any continuous distribution of betting rates whatever, the betting rate on any individual distribution will be zero. On our theory, one such distribution can be better supported than another. Hence zero betting rate does not indicate equal support for any other proposition with the same zero rate.

Yet it will still be replied that, on this point, the spirit of our theory is that of Bayes: equal support on minimal data seems an application of the principle of indifference, and this theory lies behind Bayes' theory. I completely dissent. (i) The principle of indifference is, in my opinion, not behind Bayes' theory as expounded by Bayes; anyway, no correct application of that principle implies Bayes' theory. (ii) Our theory is not an application of the principle of indifference; it is merely consistent with it. (iii) When not cast in a self-contradictory form, the principle of indifference is not a principle but a truism completely incapable of implying anything of substance. Here is how it works: on the minimal data that each of a family of propositions is possible, then, vacuously, each member of that family is equally ill supported. It is cheering that our theory of support does not contradict such a barren tautology. But this has nothing to do with the foundation of our theory; it just shows that in the case of degenerate data, our theory does not deny a merely pompous platitude.

CHAPTER XIII

# THE SUBJECTIVE THEORY

The subjective theory of statistical inference descends from Bayes; its immediate progenitors are Ramsey and de Finetti; Savage is its most recent distinguished patron. It aims at analysing rational, or at least consistent, procedures for choosing among statistical hypotheses. In some ways reaching further than our theory of support, it is in a correspondingly less stable state. But the two theories are compatible. Some tenets of subjectivists conflict with the theory of support, but do not seem essential to subjectivism. Both theories combine elements of caution and ambition, but where one is bold, the other is wary. The present chapter will not contribute to the subjective theory, but will try to distinguish its province from the theory of statistical support.

*The two theories*

As presented in this book the theory of support analyses only inferences between joint propositions; essentially, inferences from statistical data either to statistical hypotheses, or else to predictions about the outcomes of particular trials. It attempts a rather precise analysis of these inferences, and in so doing is perhaps too audacious, but at least it proffers postulates rich enough in consequences that they can be tested and, one hopes, revised in the light of counter-examples. But in another respect the theory is timid, for it gives no account of how to establish statistical data. Thus although it may claim to encompass all that is precisely analysable in statistical inference, it cannot pretend to cover all the intellectual moves made by statisticians. It would defend itself against criticism on this score by maintaining that the way in which statistical data are agreed on—partly by experiment, partly by half-arbitrary, half-judicious simplification—is not peculiar to statistics and so should not seek an especially statistical foundation.

But the subjective theory embraces all work done in statistics, and perhaps all inductive reasoning in any field. At the same time it hesitates to appraise inferences made on the basis of particular statistical data. It is concerned only that inferences be consistent with each other. The theory's key concept has been variously called 'subjective probability', 'personal probability', 'intuitive probability', 'probability' and 'personal betting rate'. The first two phrases are most common nowadays, but I shall use the last since it bears its meaning most clearly on its face. All the terms are intended as subjective ones: if a person acquainted with some facts offers a rate for betting on the occurrence of some particular event, and if he will offer none lower, then, as a first approximation, this is his personal rate for betting on that event at that time. We can of course consider corporate, or national, betting rates along the same lines.

The theory maintains that if a person is rational, or consistent in certain ways, his rates for betting will satisfy Kolmogoroff's axioms. It argues that these axioms provide a sort of standard of internal consistency for his personal rates. If he assents to the axioms he can use their logical consequences to co-ordinate his rates. In particular he can use Bayes' theorem to pass from old rates to new as he collects more data. Since the taking of decisions may be construed as the accepting of wagers, the scope of the theory is immense. Moreover, passing from old rates to new in accord with the experimental data looks like the essence of statistical reasoning: if a man's rate of betting on $h$ increases as his knowledge grows, then he must be getting more and more certain of $h$.

As so far stated, the subjective theory and that of statistical support are obviously compatible. They certainly should be. Doubtless there is some logic of betting; if a theory of support asserted that on some data, the hypothesis $h$ is better supported than $i$, while a subjective theory said that in consistency $i$ would have to be a better bet than $h$, then, aside from quirks (like $i$ being verifiable so you could collect on it, while $h$ could not be verified in human life span and so could not serve to settle any gambling accounts) we could only conclude that one or the other theory needed mending.

Although there can be no contradiction there can be wide differences in what people claim for the two theories. Some think statistics can be made sensible only in the framework of a subjective theory, while others think this could never happen. The truth of the matter is that those inferences which are peculiarly statistical need not be founded in the subjective theory, and that at present the theory should not be considered their foundation. But some future subjective theory, more refined than the present one, might turn out to be the ideal container for statistical inference, and, indeed, for any other inductive inference.

*Axioms*

The subjective theory demands a logic of consistent betting rates. If a man offers a set of rates on different options, then presumably they should satisfy a canon of consistency. It has long been contended that some form of the Kolmogoroff axioms is just what is needed. We have already seen Bayes making this very assumption. But there is a difficulty, glossed over in discussion of Bayes, which arises from the very notion of betting. If the rates satisfy the axioms, some linear scale of values seems implied. Now questions about utility and how it should be measured are, as theoretical economists know to their cost, extremely difficult to analyse aright. So it is worthwhile reciting three ways of shelving questions about value, or else of developing definitions of utility concurrently with those of betting rates.

First, following Ramsey, betting rates and utility may be defined jointly in terms of some postulates about rational choice.† I am inclined to suppose this the soundest approach to the general theoretical problems; anyway, Ramsey's methods lead direct to the usual axioms.

Secondly, following de Finetti and Savage, one can begin with mere comparisons of betting rates; my personal rate for betting on the occurrence of $E$ is at least as great as that for betting on the occurrence of $E'$ if, offered the option, '*either* bet on $E$, receiving a prize you desire if $E$ occurs, and nothing if not, *or* bet on $E'$, receiving the prize if $E'$ occurs and nothing if not', I always opt for $E$. De Finetti showed that if any betting rates on any two events can be compared, and if for any integer $n$ there is a set of $n$ mutually exclusive events between which my betting rates are indifferent, then my comparative rates can be metrized, and this metrization must satisfy Kolmogoroff's axioms.†

† F. P. Ramsey, 'Truth and probability', *The Foundations of Mathematics and other Logical Essays* (London, 1931).

† B. de Finetti, 'La prévision', *Annales de l'Institut Henri Poincaré*, VII (1937), 1–68. See also L. J. Savage, *The Foundations of Statistics* (New York, 1954), ch. III.

## AXIOMS

Thirdly, following another idea of de Finetti's, more recently exploited by logicians, a plausible consideration about betting entails the Kolmogoroff axioms.† Suppose I offer rates over any field of events. I would be very foolish if a skilful entrepreneur, simply noting my rates, could make a Dutch book against me, that is, so arrange his wagers at my rates that he was sure to win no matter what happened in the world wagered on. If I offered 3:1 on Wolves winning the next Cup Final, and 3:1 on their not winning, and was certain they would either win or not win, he could be certain of a profit no matter what happens on the football field if he bet suitable sums against me on all counts—win, lose, or draw. If I am a sensible bettor my odds will not be open to such a Dutch book; they will not even be open to a book such that I cannot profit though my opponent can. The necessary and sufficient conditions for avoiding a Dutch book are, it turns out, provided by Kolmogoroff's axioms. Ramsey cites this as a pleasant corrollary of the axioms; others have taken it as their very justification.

Thus whatever be the final analysis of utility and betting rates, there are lots of ways of arguing that reasonable rates satisfy the usual axioms. There are still many points worthy of questioning in this regard, but they will not be discussed here. The usual axioms are a powerful machine: Bayes' theorem follows at once.

### Chance and the subjective theory

De Finetti denies the very existence of the physical property I call chance; a world possessing such a property would be, he says, a 'mysterious pseudo-world'.‡ Earlier he was inclined to say much the same thing about many other physical properties. He even hinted that all the laws of logic are themselves merely the product of a coincidence in subjective inclinations. This provides an interesting metaphysic, but hardly tells much about statistics as it stands. If all flesh is grass, kings and cardinals are surely grass, but so is everyone else and we have not learned much about kings as opposed to peasants.

† J. G. Kemeny, 'Fair bets and inductive probabilities', *Journal of Symbolic Logic*, XX (1955), 263–73; and papers by A. Shimony and R. S. Lehmann in the same volume.
‡ B. de Finetti, 'Probability, philosophy and interpretation'; an article for an *International Encyclopedia of the Social Sciences*, English mimeographed version circulated by L. J. Savage (1963), p. 12.

De Finetti's work is exciting because he has a subjective analysis of what he takes to be the chance of an event. If I say the chance of heads with this coin is 1/2, then, according to de Finetti, I am trying to convey the fact that I am as ready to bet on heads as on tails, and to act in ways conformable to my willingness to bet thus. If I say that tosses of this coin are independent, and that I do not know the chance of heads, I am trying to convey the fact that, on $n$ tosses, I consider the rate for betting that exactly $k$ heads fall, in some particular order, equal to the rate for betting that they fall in any other order in the same $n$ tosses. In the framework of my betting odds any future sequence of $n$ tosses with exactly $k$ heads is, as he puts it, *exchangeable* with any other sequence.

But contrary to what de Finetti claims I was trying to say, and despite the mathematical interest of his analysis, that is not what I was trying to say at all. I was trying to say, and did say, that the coin and tossing device have a certain physical property. It is widely held that this property derives from others, like material symmetry and gravity. Perhaps if I believe tosses are independent my rates will be exchangeable, but it in no way follows from exchangeability of my betting rates that the world has those characteristics making for independent trials.

This point about the existence of a physical property of chance or frequency in the long run could be made in connexion with more recently advanced physics, where it is currently held that the physical properties of photons and other tiny entities can only be described in terms of chances. To talk about chances and unknown chances in this domain is not to talk about gambling and beliefs but about the behaviour of minutiae. Yet this kind of example only sounds impressive, for the more one masters this branch of physics, the more one must wonder exactly what it signifies. The point about chance is more clear with coins. Get down to a particular case and experiment with coin flipping in a primitive device, say shaking the coin in a can and letting it fall on the carpet. Suppose it seems persistently to fall more often heads than tails. It really requires tremendous progress in philosophy even to think the propensity to have a greater frequency of heads might not be a property of the coin and tossing device. It is hard to give an account of the property. But this, I conceive, is no ground for dismissing the property from the pantheon of subjects fit for scientific enquiry.

De Finetti has made some play with the fact that chances can be unknown. One's own betting rates, of course, cannot be unknown except in so far as one suspends a rating while one performs a computation. De Finetti would also do away with unknown chances, which, he says, 'are of course meaningless in a subjectivistic theory'. Why, one wonders, are they meaningless? Are objective melting points meaningless too? 'It can fairly be said', or so he continues, 'that an objective probability [a chance] is always unknown, although hypothetical estimates of its value are made in a not really specifiable sense.'†

To grasp this remark we had best have an example. First take the coin tossing device; suppose a lot of tosses make it seem indubitable, for whatever as yet unanalysed reason, that the chance of heads is about 0·64. Compare this with an experimental set-up which makes it seem indubitable that the melting point of some new alloy is about 986·4° C. Considering de Finetti's remark, what might be said about these two cases?

Assuming that, as asserted, we cannot ever know the chance of heads, there are two possible opinions. (1) Both the chance of heads and the melting point of the alloy are unknowable. (2) The chance of heads is unknowable, but the melting point of the alloy can be known.

The first alternative can take several forms, all pretty familiar in philosophy. (1a) We cannot 'really' know the chance of heads or the melting point, for we cannot *know* anything except a few logical truths. (1b) We can know some contingent truths, like the fact that I have a ring on one finger now, but we cannot know the truth of general propositions, especially those with a dispositional flavour.

Each version subdivides again. In one subdivision of each, it would be contended that although we cannot really know about chances, still we can approach certainty; there are objective truths, even if they are not completely knowable. In the other subdivision, it is held that there is nothing to know outside our opinions or our minds; neither melting points nor chances characterize reality at all. So (1a) and (1b) divide into stark empiricism, on the one hand, and idealism on the other.

If de Finetti has in mind some thought which can be pigeon-

† *Op. cit.* p. 12.

holed under (1), it seems, as Marxist writers already claim, to be the idealist form of (1b). This is no place to argue with such a metaphysic, for it is of no special interest to statistics. Every form of (1) contends that *all* flesh is grass, a helpful insight, but not a helpful insight about kings as opposed to peasants, statistics as opposed to metallurgy.

But if de Finetti would reject (1) and maintain (2), he must hold that situations about melting points and chances differ. What would be his ground? We must not confuse difficulties about analysis and difficulties in knowing. It is very hard to analyse how we know about melting points or, indeed, how I know I am seated before a typewriter. But this is not to say we never know. I fear de Finetti rests too much of his argument on lack of good analyses of statistical inference.

Now if you had to rely on what operationalist philosophers say about melting points, you would have no faith in melting points, but fortunately you can go out and determine the melting point of copper yourself. There are lots of philosophical questions you can raise but that does not show there is no melting point of copper. It shows that more analysis and understanding is needed. Similarly, de Finetti is right to scoff at those who too readily take for granted the principles of statistical inference, who, for instance, have glibly spoken of long runs in an infinite sequence as if this helped understand experimental work. But the best of these thinkers, such as von Mises, have begun their books with lucid appeals to common experiences with coins and life insurance, and invited their readers to analyse this experience. Toss coins, says Mises, just as you might say, melt some copper and see for yourself. I do not think Mises' analysis is correct, and future workers will not think my analysis correct. But the difference between de Finetti's attitude and that of the present essay is that he, clearly perceiving the difficulties which beset analysis of objective long run frequencies, dismisses the physical property out of hand, whereas I have tried to analyse how one does reason about the property, and have tried to bring out the principles of reasoning about it in the hope that this will define it better, and lead other people to compose definitions which are better still.

I think it may be something of an historical accident that some subjectivists distrust the physical property of chance. There is only one matter, discussed below, which could incline a theory of consistent betting rates to avoid the physical property. Hence in relating the theory I shall sometimes write as if subjectivists admitted the physical property. If any reader finds this too much of a distortion, he may try to translate my 'chance' into de Finetti's analysis by way of exchangeability. I shall be like the subjectivist Savage, who says he is usually content to write as if there were objective long run frequencies, even though he believes this is only a manner of speaking.

*Betting on hypotheses*

In one respect the purist may be right to want to get rid of chances. Here is a sort of picture of Savage's view of statistical inference. We begin with some initial betting rates appropriate to a class of statistical hypotheses; experimental data are obtained; by Bayes' theorem we compute posterior betting rates on those hypotheses. But what of the crucial matter of betting on hypotheses? What sense does it make to bet on something which cannot be conclusively verified? Can it make sense to bet on theories? De Finetti says not, and he seems correct. It is not sensible to bet unless it is conceivable that you should collect your winnings without dispute. Now if this view is accepted, my crude sketch of Savage on statistics cannot survive. For it uses bets on hypotheses, and these are now ruled out of court.

There are several putative solutions to the difficulty.

(1) Abandon betting rates as the central tool in subjective statistics, and revert to a subjective form of Jeffreys' 'probability', or degree of confidence. Even if it is foolish to bet on statistical hypotheses, one can still be more confident of one than another. A few pages ago I described some ways of deriving Kolmogoroff's axioms to get Bayes' theorem. All rely on the notion of choosing among options, or, indeed, betting on possibilities. So if we revert to Jeffreys' probability, these derivations must be rejected. But Jeffreys has other arguments for the axioms; more recent ones due to Cox may be more compelling.† However, the methods of Jeffreys and Cox leave a good deal to convenient convention, and the precise form of Bayes' theorem will stem from conventions.

† R. T. Cox, *The Algebra of Probable Inference* (Baltimore, 1961), ch. 1.

Some will find this a drawback. It destroys the pleasant behavioural background for much subjective theory which has served to make it attractive.

(2) Reject de Finetti's objection to betting on hypotheses. This seems to be Savage's view. You might even imagine yourself pretending to bet on theories and hypotheses, as if you were gambling with an omniscient being who always pays up if you are right. Evidently such an idea is very alien to de Finetti's.

(3) Adopt a very strong philosophical position, that no generalizations whatsoever are true or false; at most they codify suggestions for action. Thus the assertion, 'the mean survival time of insects fed poison $A$ is less than that for insects fed poison $B$', is not to be wagered on for it is neither true nor false. One can only bet about particular insects, say, the batch which currently infest my row of beans. I can bet about this, and use Bayesian methods to adjust my betting rates. But note that it is not merely statistical hypotheses which are abandoned: all hypotheses and generalizations must go, even 'all men are mortal'. Ramsey has sketched such a position,† and Mill's *System of Logic* tried to apply it rather more extensively.

(4) Analyse statistical hypotheses—but not all generalizations—out of existence via de Finetti's exchangeability. De Finetti's work is the prime source of this idea; a more powerful mathematical analysis has been given by others.‡ For all its elegance it would be very helpful if a subjectivist could write out, from beginning to end, a piece of statistical inference, ostensibly concerned, say, with statistical hypotheses about the mean life of insects under two different treatments. It has already been suggested that this cannot be done in full.§ Perhaps, with sufficient ingenuity, the task can be completed, but only when it is completed can we see exactly what we are being asked to accept.

These, then, are four of the different ways of overcoming doubts about the sense of betting on statistical hypotheses. I do not know which is to be preferred. In what follows I shall write as if one at

---

† F. P. Ramsey, 'General propositions and causality', *The Foundations of Mathematics* (London, 1931).

‡ E. Hewitt and L. J. Savage, 'Symmetric measures on Cartesian products', *Transactions of the American Mathematical Society*, LXXX (1955), 470–501.

§ R. B. Braithwaite, 'On unknown probabilities', *Observation and Interpretation in the Philosophy of Physics*, ed. Körner (London, 1957).

least is satisfactory, and for brevity shall write as if you can bet on statistical hypotheses. I confess that if I thought the subjective theory were the theory for founding statistics, I could not be content with this. But here I am concerned only to state the theory, and distinguish it from the theory of support.

*Learning from experience*

One of the most intriguing aspects of the subjective theory, and of Jeffreys' theory of probability, is use of a fact about Bayes' theorem to explain the possibility of learning about a set-up from a sequence of trials on it. The fact seems to explain the possibility of different persons coming to agree on the properties of the set-up, even though each individual starts with different prior betting rates and different initial data.

Consider any chance set-up and independent trials of kind $K$. Given any prior betting rate and any statistical hypothesis $h$ about the distribution of chances of outcomes on trials of kind $K$, let $R_1$ be the posterior rate for betting on $h$, based on the prior data and knowledge of the result of a first trial of kind $K$. Let $R_n$ be the posterior rate based on the prior data, and the result of a compound trial consisting of a sequence of the first $n$ trials of kind $K$. Then under very general conditions the sequence $R_1, R_2, ..., R_n, ...$ converges, no matter what the prior betting rate. Moreover, let $S_1, S_2, ..., S_n, ...$ be any other parallel sequence, based on some different prior betting rate. Then so long as neither prior betting rate was zero, $R_n$ and $S_n$ come increasingly close together. Both converge on the same limit.

It follows that if two persons have widely divergent prior rates for betting on $h$, their posterior rates will come close together; in fact this will happen after relatively few trials, for the convergence is pretty rapid. The only major exception is when one person thinks an hypothesis is impossible, while the other does not.

These elegant facts show the subjective theory is consistent with learning from experience. That is, they show it is consistent with what we all like to believe, that reasonable people will come to agree so long as they have done enough experimenting.

Of course we should not use these elegant facts as empirical support for the subjective theory. For there is at present no reason to believe that the approach to agreement among intelligent men resembles, in detail, the course charted for it by the subjective theory. At present most of the interesting researches on this subject are still occupied with trying to measure betting rates, and how to elicit them from persons without radically altering the structure of people's beliefs. Other more theoretical investigations are entangled with the notorious problems about defining utility. When these matters are cleared up it may turn out that the approach to unanimity among men of good will roughly follows the course charted by the subjective theory. But this would not really be empirical support for the theory, for the theory is normative. If people do not tend to agree as the theory says they should, this shows not that the theory is wrong, but that if the theory is right people are stupid.

*The likelihood principle*

We have seen how a man can learn from experience. Now suppose a man entertains some hypotheses about a set-up. He makes an experiment to help discover which are true. How much does he learn about the hypotheses from the experiment? The subjective theory has a plausible answer. If the experiment was profitable, he will want to change his betting rates; if he does not change his rates, then he is just where he was before, and has learned nothing about the hypotheses from his test. So the change in his betting rates is a due register of what he learns.

Or so it has been suggested. It is a good rough model of what a man learns, but is hardly very precise. For instance, as the result of his experiment, the man may leave his initial rates as they were, and yet feel very much more confident that those rates are appropriate. This is perhaps particularly plain if one is betting on what will happen at the 1000th trial on a set-up; to begin with, one may very hesitantly give even odds to heads and tails, and after 500 trials very confidently offer those odds. This change in confidence reflects learning about the experiment not directly registered in a change in the odds.

However, let us suppose that changes in betting rates are a fairly sound gauge of what has been learned. We imagine Abel and Baker entertaining some hypotheses about a set-up, and offering odds on these. Abel makes some trial on the set-up. He conceals the results from Baker, but does reveal the likelihood function induced by the results—the likelihood function being that which gives the relative likelihoods of the various hypotheses, in the light of the new data. Then each man computes his new betting rates. Both must use Bayes' theorem, which says the posterior rates are in proportion to the prior rates times the likelihoods. Hence Baker is as well off as Abel. If changes in their rates record what they have learned, Baker has learned as much as Abel. More generally, the likelihood function sums up the significance of the experiment to Abel and Baker, and indeed to anyone else.

This suggests what Savage calls the *likelihood principle*. In his words, 'Given the likelihood function in which an experiment resulted, *everything* else about the experiment...is irrelevant'.† The same sort of idea is to be found in Fisher, and in more explicit form, in some remarks of Barnard, but neither of those authors uses it as more than a rule of thumb which is handy before the foundations of statistics is better understood. Neither much liked subjective statistics, and neither would incline to the Bayesian derivation of the principle. Savage, on the other hand, would have it as a part of the very *credo* of subjective statistics. Since so much of the present essay has relied on likelihoods, it may seem as if I too should be committed to the likelihood principle. So it may be useful to state the very strongest of such principles which might be got from the law of likelihood:

(*A*) If the only prior data about a set-up consists of statistical data, namely data about the class of possible distributions and about the class of possible results on different trials, then a new experimental result on the set-up can be evaluated on the basis of this data by considering solely the relative likelihoods of the various hypotheses in the light of the experimental result.

This is not entailed by the theory of support, but is a plausible extension of it. It falls short of the likelihood principle in two distinct ways. First it is concerned solely with what I have been calling statistical data. There might be other facts about the set-up, perhaps involving important patterns of symmetry, which would be useful for evaluating experimental results. A plausible, though by no means certain, example of this sort of thing has just been

† L. J. Savage, 'The foundations of statistics reconsidered', *Proceedings of the Fourth Berkeley Symposium on Mathematical Statistics and Probability* (Berkeley, 1961), I, p. 583.

published.† If such examples are sound, the likelihood principle is refuted, while $(A)$ still stands. In any event, $(A)$ does not entail the likelihood principle.

Even more important, $(A)$ is about how experimental results are to be appraised when some 'statistical data' are truly taken as datum. But as in any discipline whatsoever, it is quite conceivable that what was earlier taken for granted should, in the light of experiment, turn out to be quite doubtful. An experimental result may force us to be sceptical of our original statistical data. $(A)$ does not entail that when this happens likelihoods are the sole guide for inference. They may be, of course, but $(A)$ does not entail it. So once again the likelihood principle makes stronger claims than $(A)$.

So two separate questions arise: is the likelihood principle true? And, does it have any useful consequences which are not consequences of $(A)$? We need not be concerned with the second if the first can be definitively answered. But at present I do not think it is known whether the likelihood principle is true or false. Birnbaum has derived it from two other principles, one of which is a plausible assertion about the irrelevance of, for instance, the way in which experiments are combined, while the other is a principle stating that sufficient statistics give as complete a report of the significance of an experiment as does the actual experimental result itself.‡ Unfortunately when stated in full generality his principle of sufficiency is no more certain than the likelihood principle. It could be expressed in a weaker form, perhaps, but then it would entail no more than $(A)$.

Part of the difficulty in trying to comprehend the likelihood principle is the vagueness of such terms as the 'significance of the experiment'. Birnbaum introduces the expression 'evidential meaning' and even has a symbol which stands for it. This is fair enough for deriving the likelihood principle from the sufficiency principle, where 'Ev( )' is an uninterpreted term in the formal statement of each, but it does not help when one is trying to discover whether some interpretation of these principles is true.

† D. A. S. Fraser, 'On the sufficiency and likelihood principles', *Journal of the American Statistical Association*, LVIII (1963), 641–7.

‡ A. Birnbaum, 'On the foundations of statistical inference', *Journal of the American Statistical Association*, LVII (1962), 269–306.

One must, of course, use undefined terms in any analysis. I have taken 'support' without definition throughout this book. But 'support' differs from 'evidential meaning' in two respects. First, it is customary in English to speak of evidence supporting hypotheses. So one may to some extent rely on this familiar use of the English term 'support'. But although one can in some contexts make sense of 'evidential meaning' through more familiar locutions like, 'this new evidence means you were right all along', there is no ready-made use for the expression 'evidential meaning'. So one cannot rely on some established practice in trying to understand Birnbaum's statement of the likelihood principle. Moreover, although 'support' has meaning in English, it has further been possible to characterize a logic of support as represented, for instance, by Koopman's axioms. This has not yet been done for 'evidential meaning'. If it were done, one could be more certain of what is supposed to follow from the likelihood principle. Until it is done I do not think we can say with Savage that the fact that some form of the likelihood principle 'flows directly from Bayes' theorem and the concept of subjective probability...is...a good example of the fertility of these ideas'.†

*The unexpected hypothesis*

It has sometimes been suggested that if the likelihood principle is true, an experimenter need only publish the likelihood function induced by his experiment. This opinion seems mistaken. The likelihood function induced by the experimental results gives, in the light of the data, the relative likelihoods among various hypotheses. But what hypotheses? Call the *complete* likelihood function one which can take as argument any statistical hypothesis whatsoever. In any actual case such a function will be literally unprintable. For it is not even clear, from the standpoint of set theory, whether there is a set of all statistical hypotheses; but suppose there is such a set; since the likelihood function over it will take no recognizable analytic form, it will not be possible to state a complete likelihood function. Hence the experimenter can only write down a likelihood function defined over some small class of hypotheses. This may be a very natural class, perhaps the class

† L. J. Savage and others, *The Foundations of Statistical Inference* (London, 1962), p. 17.

of all hypotheses he thinks possible. But it is always conceivable that the results themselves cast doubt upon, or even shatter, his earlier assumptions. Or he may publish not realizing that such doubt is cast: a hitherto unexpected hypothesis may occur to someone else. From the limited likelihood function which he publishes other workers will not in general be able to evaluate hypotheses which the original experimenter did not consider. Hence no printable likelihood function will ever convey as much as the full experimental data would have done.

It is instructive to look more closely at unexpected hypotheses. Imagine the coin tossing experiment, in which it is supposed that tosses are independent. If someone were interested in the distribution of chances of outcomes on trials consisting of 100 tosses, he would probably take it for granted that individual tosses are independent. Coins are like that. So he assumes his problem is one of estimating the chance of heads at any single toss; and from that he could compute the distribution of chances for trials consisting of 100 tosses.

He tosses a while and begins to notice that on any assumption of independence, his results are very odd indeed. He begins to doubt; further experiments demolish his former assumption. Whenever the coin falls heads thrice in succession after a tail, it falls heads a fourth time. There are no runs of exactly three heads. So he is forced to conclude tosses are not independent.

My theory of statistical support does not attempt rigorous analysis of the reasoning here. It can only tell what is well supported by various pieces of statistical data; in the above case a man is forced to reject what once he took for datum. Many considerations affect him. One is the cogent and rational desire for simplicity. If he suspects a little dependence among his tosses he will probably ignore it and essentially deny that it is there. The rational inclination to make simplifying assumptions is one of the factors which influence acceptance of statistical data. The theory of statistical support cannot judge the force with which an experiment counts against a simplifying assumption. Nor is this a matter of mere pragmatics, irrelevant to questions of support, truth, and justified belief. The relation between simplicity and truth is an unknown quantity, but is not nil. Since the relation does not seem peculiar to statistics, it is not treated within the theory of statistical support, but this is not to deny its existence.

The subjective theory, on the other hand, is not formally prevented from examining the case of the unexpected hypothesis. Here is a point at which one might hope that the subjective theory could give real help to philosophical analysis. The theory would say, I suppose, that the initial betting rate on hypotheses assuming dependence is very low. Some trials indicate dependence; this shows up in the likelihood ratios. Finally the likelihoods come to dominate the initial rates. The unexpected, namely what had a low betting rate, is now thought probable, namely has a high betting rate.

It is crucial that the initial rate was not zero, for no amount of new data can elevate a zero betting rate. Not, that is, according to the subjectivist use of Bayes' theorem. In a similar case, concerned with the alleged Normalcy of a sample, Savage writes that after abnormal data are garnered, 'all [persons] concerned have (and presumably always had latent) doubts about the rigorous normality of the sample'.† He is forced to say this because, on his theory, if you come to doubt, you must always have had a little doubt shown by the original betting rate being a little above zero. But is what he says true?

Perhaps Savage intended a mere pleonasm; perhaps what he meant by a 'latent doubt' is what someone has, by definition, if he ever comes to doubt something. Then it is entirely empty to say that if a man doubts, he must always have had latent doubt. But if what Savage says means that if you come to doubt something you must in consistency always have had a tiny shred of doubt, then what he says seems false. Yet this interpretation on his words is required by his universal use of Bayes' theorem.

It seems plain that a man can be completely certain of some proposition, yet later, as he learns new evidence, come to doubt it. This is a veritable mark of rationality. It is reasonable, first, to be certain of some things; a man really uncertain of everything would be a fool. It is reasonable, secondly, to take account of evidence, even retracting what one once thought certain; a man who does not is a bigot. So it can be reasonable for a man to be completely certain of something and later come to doubt it. In particular, it is possible that a man should be completely certain that tosses with a

† *Op. cit.* p. 16.

coin are as near to independent as could possibly matter, and yet, after tossing many times, be forced to conclude that tosses are markedly dependent.

It may be protested that whatever the man's certainty, his betting odds on dependence will not be strictly 0. Here we have a difficulty with betting odds. I am certain, say, that I now have a ring on a finger of my left hand. But I would not offer odds of infinity to zero on this event, because it does not make sense; it only makes sense to go through the ritual of betting if there is a prospect of gain. But I shall offer as gigantic odds as make sense; I am cheerful to stake my fortune on my claim, for the prospect of winning some trifle, say 10 shillings. Indeed if a shady fellow were to approach me now proffering these odds, a fellow with a diabolic or at least hypnotic look about him, I should hesitate, because the very fact that he is prepared to make such an offer is itself evidence that something is fishy, and may itself rid me of my former absolute certainty. But if I am betting with a harmless statistician, and if I can choose the object to be wagered on, I shall offer as great odds as can make sense. I can see no reason, if we are to make some ideal, conceptual, version of betting odds, why it should not be said that my betting rate is zero to infinity. And if I do offer these odds, but later learn of such facts as force me to retract them —I learn I was under mild hypnosis, and in reality my ring had been filched—there seems nothing in logic or the nature of betting odds to say I have been inconsistent.

But here we are losing our way; at its extreme ends the clear path of betting rates becomes a labyrinth of poor analogies. Contrary to the evidence, let us suppose that Savage is right: if ever I become unconvinced, then, we suppose, I am inconsistent unless my earlier betting rates differed from zero. But how does this work in practice? In the coin tossing case, I have various hypotheses I think possible; all assume independence. My betting rates on all these cannot add up to 1, for there must be other hypotheses on which my rate is not zero. A whole infinity of these; worse, an infinity of very high order. Barnard put the point admirably in discussion: 'Savage says in effect, "add at the bottom of the list $H_1, H_2, \ldots$ 'something else'"'. But what is the probability that a penny comes up heads given the hypothesis "something else"? We do not know. What one requires for this purpose is not just that there should be some hypotheses, but that they should enable you to compute probabilities for the data, and that requires very well defined hypotheses.'† Barnard has in mind the fact that use of Bayes' theorem requires multiplying a likelihood by a prior betting rate; the likelihood of 'something else' is not defined. If it is not defined, the subjective theory does not adequately treat the case of the unexpected hypothesis. It is a shame that where one looked for profit, no comfort is to be found.

There remains a prospect of comfort. Once again Jeffreys' analysis, with its greater attention to philosophic detail, has foreseen the problem. As remarked above, one of the considerations which first militates against assuming non-independence is that independence is so much simpler. Jeffreys supposes that possible hypotheses could be arranged in some scale of simplicity. Then one would assign high probability, and a high betting rate, to the simplest hypotheses. But there would be clearly defined betting rates over less simple hypotheses; slight rates indeed, which would vanish as the hypotheses became too complex for human analysis.‡ This is a tremendously promising suggestion. Jeffreys sketches how he thinks it should go, though he does not seem to apply his suggestion much. But all we need here is a scheme of the method. Unfortunately it is not at all clear that his analysis of simplicity is satisfactory. More recent work on the subject suggests that every plausible canon of simplicity breaks down.§ It may, however, turn out that although there is no universal measure of simplicity, there are still particular and rather subjective measures which individuals could adopt in special problems, and which would underlie their reaction to unexpected hypotheses. It is to be hoped that a philosophically minded subjectivist will explore Jeffreys' proposal and see if it can be adapted in a rigorous manner.

*Conclusion*

It seems from this survey of some tenets of subjectivism that there can be no inconsistency between, on the one hand, the theory of statistical support, as an analysis of chance and a foundation for

† *Op. cit.* p. 82.
‡ H. Jeffreys, *Theory of Probability* (Oxford, 1948), pp. 47 ff.
§ R. Ackermann, 'A neglected proposal concerning simplicity', *Philosophy of Science*, XXX (1963), 228–35.

the statistical inference to which it pertains, and, on the other hand, the subjective theory as an analysis of consistent inference. The subjective theory will certainly have practical applications alien to the present theory of support. To state only the obvious, there may be an array of hypotheses about which there is some not easily evaluated data, but which does give some reason to prefer some hypotheses to others; then it may be desired to make further, precisely analysable, experiments, and to programme a computing machine to appraise the hypotheses in the light of both new and old data; the computing machine cannot make human decisions, so it should be fed some prior distribution of betting rates, based on the old data, and complete its evaluation by use of Bayes' theorem. Some idea of the scope of such activities can be gleaned from recent publications.†

But from the bias of the present volume, it is important to know whether the subjective theory contributes to an understanding of that physical property, chance, or to rigorous analysis of the logical inferences which can be made about it. Some day the subjective theory may be in a position to contribute to this philosophical task, but, I think, not yet. Many subjectivists, and especially de Finetti, have, of course, not wanted to do anything of the sort, for they have denied the very existence of the physical property analysed in this essay. But there is nothing in a theory of consistency to deny long run frequency; denial of the property is an optional extra, which, one hopes, may not long remain in fashion.

At one time, in what might be called the era of conceptual parsimony, it was customary for students of probability to feud among themselves as to which concepts are really needed for their analyses. Some abhorred not only subjective betting rates but even notions like support; there were others who thought the very idea of long run frequency absurd. Such quarrels reflected what was a commonplace among logicians of the time, that it is desirable to use as few different concepts as possible; this, it seemed, can drive out inconsistency. Nowadays, precisely because of the work of earlier logicians, it is plain that frugality in logic has no intrinsic merit. One uses whatever tools are convenient, ensuring, as one goes along, that there is a respectable underlying logic with

† R. Schlaiffer, *Probability and Statistics for Business Decisions* (New York, 1959).

which one can prove the occasional consistency theorem. Thus at one time a subjectivist might have been gratified by my extensive use of support in analysing chance, for support has some vague kinship with subjective betting rates. My use of the concept of support certainly marks a difference between my analysis and that of the early frequentists. But this does not reflect the triumph of subjective ideas, but only the passing of parsimony. It can also reflect the end of exaggeration, for nowadays it would be absurd to contend that long run frequency is all there is to probability.

Neither frequentists nor subjectivists have been right about probability, but to discover exactly what there is to probability one needs a very different type of analysis than anything found in this essay. Frequency in the long run is a property which men have only begun to notice and use as a tool in inference. Our task has been to state and test postulates which serve to circumscribe, and so to define, the property which can be discerned in a few well known examples. We could not pick out the property by its name in English because there is no standard name for it. Probability, on the other hand, is more established: nearly every adult speaker of English can use the word 'probable' correctly. No account of probability could be sound if it did not examine what the English word 'probable' means. Some say the word and its cognates are entirely vague, but our ability to use them regularly suggests they must be governed by pretty stern regularities, which mark out the concept or concepts of probability. Testing conjectures about these regularities requires new methods. Unneeded in the present study, they will be used extensively in another work.

# INDEX

*The number attached to a much-used technical term indicates the page where the term is first introduced*

Ackermann, R., 225
action and belief, 164–70
acts of guessing, 168
admissible estimates, 177-9
after-trial betting, 95-102
algebra
  Boolean, 32, 134
  sigma-, 32, 134
Anscombe, F. J., 103
antisymmetry (in the logic of support), 33 f.
Arbuthnot, J., 75–81
axioms
  absolute support (Kolmogoroff), 134 f.
  betting rates (Kolmogoroff), 193, 210 f.
  chances (Kolmogoroff), 18 f.
  relative support (Koopman), 32 f.

Barnard, G.
  on kinds of trial, 88
  on likelihood, 58
  on the likelihood principle, 65, 219
  on subjective statistics, 224
Bayes, T., 66, 112, 188, 190–227 *passim*
  definition of probability, 191 f.
  assumptions about betting rates, 192-5
  postulate of, 198–200
  before trial betting, 95–102
belief and action, 164–70
belief-guesses, belief-estimates, 168 f.
Bernouilli, Daniel, 63 f., 176
Bernouilli, James, 23
Bernouilli, Nicholas, 77
betting, before- and after-trial, 95–102
  on hypotheses, 215–17
  rates: axioms for, 210 f. and 193–220 *passim*; Bayes' use of, 192-5
Birnbaum, A., 65, 110, 220
Boolean algebra, 32, 134
Box, G. E. P., 206
Braithwaite, R. B., 10, 50, 74, 114, 216
  theory of chance, 114–17
Brillinger, D., 150 n., 151 n.
Brown, G. Spencer, 132

Cargile, J., 121
Carnap, R., 166–8, 171, 187
chance, 10
conditional, 19 f.
chance process, 15, 26
chance set up, 13
  tandem, 195 f.
chrysanthemum experiment, 88
Church, A., 119
collective (von Mises), 5
completeness
  of fiducial theory, 154 f.
  of postulates, 35
compound trials, 21–3
conditional chance, 19 f.
  trials, 14, 19 f.
confidence intervals, 159 f.
consistency of fiducial theory, 150
consistent estimators, 184
Cournot, A. A., 5, 13
Cox, D. R., 206 n.
Cox, R. T., 134 n., 215
critical likelihood ratio, 89, 111–13
cumulative distribution function, 17 f.
cumulative graph, 16

decision theory, 29–32, 75, 164, 186–8
definition, 4 f.
Dempster, A. P., 103, 151 n.
density function, 18
  experimental, 68–70, 148 f.
Derham, W., 77 n.
distribution, 16
  binomial, 23 f.
  cumulative, 17 f.
  Normal, 71, 156–8, 182–6
Dürrenmatt, F., 32
Dutch book, 211

efficiency of estimates, 185
Ellis, L., 5
epidemics, 1, 9, 86 f.
epistemic utility, 31
errors, theory of, 155, 175
estimates and estimators, 28 f., 62, 161–89 *passim*

estimates and estimators (*cont.*)
  admissible, 177–9
  belief-estimates, 168 f.
  consistent, 184 f.
  fiducial argument for, 181–7
  invariance of, 172
  maximum likelihood, 49, 62–4, 126, 176, 185 f.
  minimax, 186 f.
  minimum variance, 183 f.
  Savage's criterion for, 179–81
  scales for, 171–3
  unbiased, 182 f.
  uniformly better, 177–9
exchangeable events, 212
expectation, 30, 102
experimental density, 68–70, 148 f.

fair stake, 192
Feller, W., 9
fiducial argument
  application: to confidence intervals, 159 f.; to Jeffreys' theory, 147 f., 204; to Normal distribution, 156–8; to theory of errors, 155
  completeness of 154 f.
  consistency of, 150; inconsistencies in Fisher's form of, 151
  examples of, 136–9, 143–5
  origin of, 133
  postulates for: frequency principle, 135; principle of irrelevance, 145, 149
  structure of, 139 f.
Finetti, B. de, 208–25 *passim*
  on axioms for betting rates, 210
  on betting about hypotheses, 215–17
  on chance, 211–15
Fisher, R. A.
  on Bayes, 200
  on fiducial argument, 133–60
  on hypothetical infinite populations, 7, 13, 25, 122
  on likelihood, 56 f., 62, 219
  on invariance in estimation, 173
  on maximum likelihood estimates, 49, 176 f., 184–6
  on sampling, 130
  on sufficient statistics, 80, 110
  on testing hypotheses, 80–3
Fraser, D. A. S., 220 n.
Frege, G., 151, 201
frequency principle, 135

Fundamental Lemma of Neyman and Pearson, 93

Galton, F., 72
gambling system, impossibility of, 22
Gauss, K. F., 25, 72, 175 f., 186
Gödel, K., 36
Good, I. J., 110 n.
Goodman, N., 41 f.
Gossett, W. S., 72, 82 f., 113
s'Gravesande, W. J., 77
guesses and acts of guessing, 168–70

Halmos, P., 70 n., 110 n.
Hewitt, E., 216 n.
histogram, 16
Hume, D., 52
hypothesis
  statistical, 27
  unexpected, 221, 225
hypothetical infinite population, 7, 13, 25 f., 122

independence, 20–3
induction, 52, 125 f.
initial support, 146
invariance of estimators, 172
invariant tests, 97–9
irrelevance, 141–60
  principle of, 145, 149

Jeffreys, H.,
  on axioms for probability, 38, 134, 215
  on chance, 11
  on the fiducial argument, 140, 147
  on initial probabilities, 147, 152–4, 201–7
  on Neyman–Pearson theory, 103
  on simplicity, 225
joint proposition, 56–8
  simple, 57

Kendall, M. G.
  on D. Bernouilli, 64
  on likelihood ratio tests, 92 n.
  on minimum variance estimators, 183 f.
  on random sampling, 130–2
Kneale, W. C., 11
Kolmogoroff's axioms
  for absolute support, 134 f.
  for betting rates, 193, 210 f.
  for chance, 8, 18 f.
Koopman, B. O., 32–4

INDEX 231

Laplace, P.S. de, 25, 76, 175 f., 200, 203
law of likelihood
   continuous case, 70 f.
   discrete case, 59, 62
learning from experience, 217 f.
least squares, method of, 175
Lehmann, E.L., 92 n., 186 n.
Lexis, W., 72
likelihood, 56–8
   given data, 61
likelihood function, 148 f., 221 f.
likelihood principle, 65 f., 218–21
likelihood ratio, 70
   critical, 89, 111–13
   given data, 71, 149
likelihood ratio tests, 91 f.
likelihood tests, 89–114
   critical ratio of, 91, 111–13
   objections to, 110
Lindley, D.V., 112 n., 151 n., 155
logic, the rôle of, 34–8
   underlying, 34 f., 225 f.
Loève, M., 2 f.
long run justification, 39–52
long run rule, 40–5

Maximum likelihood, 49, 62–4, 126, 176, 185 f.
Menger, K., 15 n.
Méré, Chevalier de, 75
Mill, J.S., 126
minimax argument for long run rule, 43–5
   in decision theory, 49 f.
   in estimation, 183 f.
minimum variance estimates, 183 f.
Mises, R. von, 5–7, 9, 13, 21, 25, 214
misleading samples, 126–32
mixed strategy, 49 f.
mixed tests, 93–9
models, 7–9, 35
Moivre, A. de, 71, 77
moments, method of, 175
Montmort, P.R. de, 77

Neumann, J. von, 102
Neurath, O., 37
Newton, I., 23
Neyman, J.
   on confidence intervals, 159 f.
   on fundamental probability sets, 13
   on rival hypotheses, 81

Neyman, J. and Pearson, E.S.
   Fundamental Lemma of, 93
   on likelihood ratio tests, 91 f.
   theory of testing, 92–111; domain of, 99–101; rationale for, 103 f.
Normal distribution, 71, 156–8, 182–6

optional stopping, 107–9
outcome (of a trial) 14
   sure, 15

partition of data, 143
Pearson, E.S., 25, 83, 103, 113; *See* Neyman and Pearson
Pearson, K., 25, 72, 175
Peirce, C.S., 47, 123
personal betting rates, 209–25
   axioms for, 210 f.
personal probability, 209
pivotal quantities, 140
pivotal trials, 140
Poisson, S.D., 85
Polya, G., 9
Popper, K., 10, 14, 121
populations
   hypothetical infinite, 7, 13, 25 f., 122
   open and closed, 122
postulational definition, 4 f.
power of a test, 92–9
predicate, projectible, 41
*Principia Mathematica*, 36, 205
principle
   frequency principle, 135
   of indifference, 51, 147, 201, 207
   of irrelevance, 145, 149
   likelihood principle, 65 f., 218, 221
   of sufficiency, 110
'probability', 2, 15, 28, 136, 209, 225
probability theory, 7
process, chance or stochastic, 15
projectible predicate, 41
proposition, joint, 57
pseudo-distribution, 152–4, 204
Quetelet, L.A.J., 72
Quine, W. v. O., 15

Radon–Nikodym theorem, 70 n., 110 n.
Ramsey, F.P., 208–17
random, 2, 7, 119
random sampling, 119–32 *passim*
random sampling numbers, 129–32
random sequence, 120 f.
random trials, 119 f.

random variables, 15
refutation, 113 f.
Reichenbach, H., 6, 40, 51
result (of a trial), 13
Russell, B., 36, 151, 201, 205

Salmon, W. C., 51
samples, misleading, 126–32
sampling, random, 119–32 *passim*
  with and without replacement, 122–4
Savage, L. J.
  on axioms for betting rates, 210
  on estimation, 167, 179–81
  on likelihood, 70
  on the likelihood principle, 65 f., 218–21
  on sufficient statistics, 110
  on unexpected hypothesis, 223
scales for estimation, 171–3
Schlaiffer, R., 226
sequence, random, 120 f.
sequential test, 106
sigma-algebra, 32, 134
simple joint proposition, 57
simplicity, 85, 222, 225
size of a test, 92–9
Smith, B. Babington, 130 f.
specification of a problem, 59, 84
stake, fair, 192
statistic, sufficient, 81, 110
statistical data, 56–9, 84, 208
statistical hypothesis, 27
stochastic process, 15
strategy, mixed, 49 f.
stringency of tests, 75, 111–13
Stuart, A. S., 57 n. 92 n., 131 n., 183 n.
subjective v. objective betting rates, 193 f.
subjective theory on statistics
  axioms for, 210 f.
  compared with logic of support, 208–10
  learning from experience in, 217–18
  likelihood principle in, 218–21
  origin of, 208
  unexpected hypotheses, 221–5
  views on chance, 211–15
sufficient statistics, 81, 110
support
  absolute, 133–5
  initial, 146 f., 203–6
  relative, 27–38

sure outcome, 15
Süssmilch, J. P., 77 n.

$t$-test, 82
tandem chance set ups, 195 f.
tests, statistical
  Arbuthnot's, 75–80
  likelihood, 89–114 *passim*
  likelihood ratio, 91 f.
  mixed, 93–9
  Neyman–Pearson, 92–111
  of rival hypotheses, 79–83
  sequential, 106 f.
  size and power of, 92–9
  stringency of, 75, 111–3
  unbiased and invariant, 97–9
  uniformly most powerful, 94
  worse than useless, 99
textual critic's hypothesis, 54
Tippett, T., 129
trial, 13
  conditional, 14, 19 f.
  independent, 20–3
  kind of, 13, 87 f.
  random, 119 f.
Tukey, J. W., 84 n.

unbiased estimators, 182 f., tests, 97
underlying logic, 34 f., 225
unexpected hypothesis, 221, 225
uniformly better estimates, 177–9
uniformly most powerful tests, 94–9
  invariant and unbiased tests, 97–9
unique case rule, 40
urn models, 9, 11, 86
  example of three hypotheses, 48–52, 60, 169
utility, 28, 30 f., 192, 202, 210, 218
  epistemic, 31

variables, random, 15
Venn, J., 5, 13, 25

waiting time, 17, 66, 152 f.
Wald, A.
  on decision theory, 29, 102
  on maximum likelihood, 185
  on minimax estimation, 186 f.
  on sequential tests, 107
Whitehead, A. N., 36, 201, 205
Williams, R. M., 151
'worse than useless' tests, 99
Wright, G. H. von, 120